Management of Radioactive Waste

Radioactive Risk Set

coordinated by
Jean-Claude Amiard

Volume 5

Management of Radioactive Waste

Jean-Claude Amiard

iSTE

WILEY

First published 2021 in Great Britain and the United States by ISTE Ltd and John Wiley & Sons, Inc.

ISTE Ltd
27-37 St George's Road
London SW19 4EU
UK

www.iste.co.uk

John Wiley & Sons, Inc.
111 River Street
Hoboken, NJ 07030
USA

www.wiley.com

Library of Congress Control Number: 2021940534

British Library Cataloguing-in-Publication Data
A CIP record for this book is available from the British Library
ISBN 978-1-78630-722-4

MIX
Paper from responsible sources
FSC www.fsc.org FSC® C013604

Contents

Preface

The use of nuclear energy for military or civilian purposes inevitably leads to the production of radioactive waste. The management of this waste is one of the most serious problems facing industrialized nations.

As with all other wastes, radioactive waste can be disposed of in one of two ways: dilution or containment. A third method exists for radioactive waste with a very short physical life, less than 100 days, which is to wait, under safe conditions, for natural physical decay.

Dilution consists of reducing the radioactive risk by dispersing the radionuclides in vast compartments of the environment such as the lithosphere, the atmosphere or the hydrosphere. This can only be done for very low-level radioactive waste, even though it has been practiced more widely in the past.

Containment consists of immobilizing the waste as long as it remains radioactive. This is relatively easy for short-lived radionuclides, i.e. with a physical half-life of less than 30 years. On the contrary, it is much more difficult to ensure for long-lived radionuclides, for some of which the physical half-life is counted in millions of years. Currently, the only realistic and practicable solution found is the multiplication of physical barriers between the radioactive waste and the environment and the biosphere, the last barrier being geologically stable and impermeable layers of the lithosphere.

The classification of radioactive waste has been the subject of IAEA recommendations, but this has not prevented the multiplication of classifications in different states, which complicates possible comparisons. These classifications are based on a combination of two parameters: the waste's level of activity and the half-life of the radionuclides constituting the waste.

A major difference in classification divides nations into two categories depending on whether they practice an open or closed nuclear fuel cycle. In the latter case, a portion of the radioactive waste is removed from this classification and is considered as usable nuclear material. However, the number of states using the closed cycle is steadily decreasing, which makes it necessary to review the quantities of radioactive waste to be actually managed.

The management of radioactive waste is specific to each state. The majority of nations manage short-lived radioactive waste in surface storage facilities and a minority in underground facilities.

On the contrary, for long-lived radioactive waste, few states have definitive solutions. This is due to the fact that the containment of the radionuclide must be guaranteed for thousands of years. For low-level waste, most countries opt for dry interim storage. For intermediate- and high-level waste, the solution generally envisaged is deep geological disposal, with some countries favoring deep geological drilling.

In the field of radioactive waste management, research is very active and innovations are numerous. This does not prevent gaps in our knowledge, uncertainties about the nature of the disposal to be adopted for certain categories of waste and often a negative opinion of the public to the proposed solutions.

June 2021

Acknowledgments

Jean-Claude Zerbib, former CEA Senior Expert, radiation protection engineer, had the difficult task of proofreading, annotating and criticizing this manuscript. He also provided me with precious documents to complete the abundant literature that was used in the writing of this book. I would like to thank him very much for this.

Professor Philip Rainbow (former Keeper of Zoology, Natural History Museum, London, United Kingdom) has done the same for the English version. I warmly thank them both for their time and efforts.

I would also like to thank the members of the Scientific Council of the ANCCLI who helped me in the understanding of certain subjects. The same goes for all the members of the *Groupe Radioécologie Nord Cotentin* (GRNC), a pluralist group, for the remarkable work done together and with every courtesy.

1

Classifications and Origins of Radioactive Waste

1.1. Introduction

Compared to other categories of waste, the quantity of radioactive waste is relatively small. In France, nuclear waste represents 2 kg per year per inhabitant [AND 17a], compared to 580 kg of household waste, 900 kg of non-construction waste and 3.4 tons of industrial waste [ADE 20]. But these residues represent an immense problem because some of them are extremely radioactive and remain harmful over excessively long time scales, for some hundreds of thousands or millions of years, that humanity cannot control.

What can we do with this radioactive waste? In the past, the ocean has served as a dumping ground for nuclear powers, which have immersed tens of thousands of radioactive drums. This time is fortunately over. Some eccentric people have suggested dropping them into space. Fortunately, the idea was not pursued. The solution now being considered for the most dangerous waste is to bury it in deep layers of clay, granite, salt or tuff, hoping that nature and geology will compensate for the weaknesses of human technology [AMI 13]. Sweden was the first nation to choose an underground storage site. All other countries, faced with the concerns of their populations and the vagaries of political changes, have postponed their decisions. On the contrary, in the United States, the suspension of the Yucca Mountain storage project in Nevada, which was ready to open, is a sign of the American administration's desire to listen to the public. However, the State must find a new solution.

Since no alternative solution is yet mature, we must take our time in making a decision that will commit humanity for a long time. France, like Canada, Switzerland and Japan, has made the principle of reversibility central to its doctrine. On the contrary, Sweden and Finland do not require it, and the United Kingdom is still considering it. It is not only a question of being able to recover radioactive packages, but of leaving the decision-making process open and giving it back to the political institutions. Parliament has once again become the master of nuclear waste management and future generations have the guarantee that nothing will be decided inescapably. The approach is virtuous. Let us hope that it is not an admission of powerlessness in the face of an insoluble puzzle [AMI 13]. It should also be emphasized that this postponement amounts in practice to leaving to future generations the care to manage and pay for the waste produced by the present generation.

Those responsible for the civilian and especially the military use of nuclear energy have in the past been very unaware of the seriousness with which the problem of nuclear waste is treated today. For example, the Hanford site in the United States was heavily polluted by unauthorized dumping during intensive plutonium production after World War II. Recently, six underground tanks leaked. In the former Soviet Union (USSR), waste in the form of highly active liquid solutions was injected directly into deep storage [MAC 96]. The United Kingdom in particular, but also other countries, and even France, have thrown drums of waste into international waters, a practice that is now prohibited [CAS 02].

Nuclear energy has been questioned almost since its inception and one of the main problems concerning its social acceptability in the world is the management of nuclear waste [ROD 17]. It is therefore imperative that nuclear nations manage radioactive waste in an exemplary way.

1.2. What is radioactive waste?

A few definitions should be kept in mind. Radioactive waste is radioactive material for which no further use is planned or envisaged. Ultimately radioactive waste is radioactive waste that can no longer be treated under current technical and economic conditions, in particular by extracting its recoverable part or by reducing its polluting or dangerous nature (French Environmental Code, article L 542.1-1). Conversely, if a radioactive material also contains radionuclides, it has a potential future use. This is the case for depleted uranium or spent nuclear fuel that can eventually be reused.

A radioactive substance is a substance that contains radionuclides, natural or artificial, whose activity or concentration justifies radiation protection control. The radionuclides contained in radioactive waste can be of artificial origin, such as cesium-137, or natural origin, such as radium-226.

Radioactive waste has three main characteristics, the type of radionuclide, the activity and the half-life. The type of radionuclide contained is related to the radiation emitted (alpha, beta, gamma). The activity is the number of atomic nuclei that spontaneously disintegrate per unit of time; it is expressed in becquerels (Bq). The half-life is the time required for the activity of a radionuclide in a sample to decrease by half [IRS 13a, IRS 13b].

1.3. Classifications of nuclear waste

Waste classification is not unique. Indeed, while the IAEA has provided broad guidelines for defining and classifying radioactive waste, each state is free to use its own nomenclature.

1.3.1. *General information on the classification of radioactive waste*

As regards the classification of radioactive waste, there are two main approaches: one by a waste management channel and the other by a waste production channel. The latter approach is partly inherited from the historical concept of radiation protection.

The management pathway approach often combines the activity and lifetime parameters of the radionuclides constituting the waste. This classification was recommended by the IAEA in the Joint Convention on the Safety of Spent Fuel Management and on the Safety of Radioactive Waste Management. This classification is used in France, Belgium and Spain. Sometimes this approach is based only on activity. In Canada, for example, there are only three main categories of radioactive waste (ILW, HLW and spent fuel), except for the specific management of waste from mines. In the Netherlands, the classification has a larger number of categories, but no distinction is made between short- and long-lived waste and consequently there are no plans for surface disposal. In Germany, the classification is based mainly on the exothermic character of the waste.

The production chain approach leads to a more complex classification, with specific chains for certain types of waste, and combining activity and lifespan. This is the approach of the United States, Japan and Sweden (in fact in Sweden, the two types of approach coexist). In Finland, a category is sometimes added for waste from hospitals, universities, etc.

There are also national specificities, as in Belgium, which treats 50% of the radium sources used in the world (the result of uranium mining in the Congo, which is historically Belgian), or in Canada, which has large uranium mines. Similarly, in France, it should be noted that there is no release threshold for waste containing, or likely to contain, only very small quantities of radioactive elements [AMI 13].

1.3.2. *The IAEA's recommendations*

The IAEA proposes dividing radioactive waste into five categories, in addition to the category of waste considered as released (EW, Exempt Waste), according to two criteria, the amount of activity and the half-life of the radionuclide (Figure 1.1). These categories are very short-lived waste (VSLW), very low-level waste (VLLW), low-level waste (LLW), intermediate-level waste (ILW) and high-level waste (HLW) [IAE 09a].

In certain circumstances, such as acceptance into a radioactive waste disposal facility, Waste Acceptance Criteria (WACs) may be established for certain radionuclides. WACs are quantitative or qualitative criteria that may include, for example, restrictions on the activity concentration or total activity of particular radionuclides (or types of radionuclides) in the waste, or requirements regarding the form or packaging of the waste.

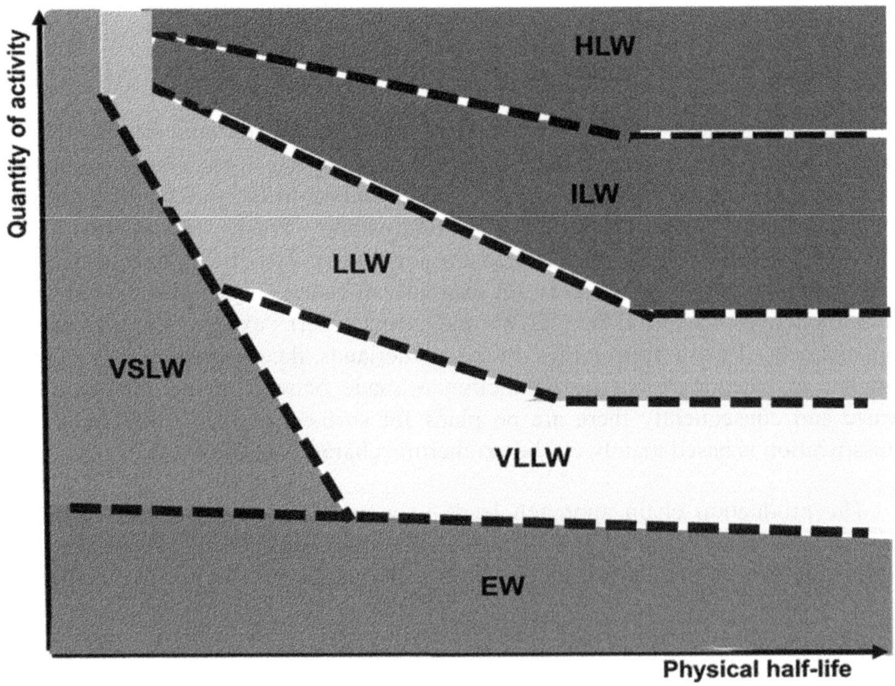

Figure 1.1. *Proposed IAEA classification of radioactive waste (source: [IAE 09a]). EW: exempt waste; HLW: high-level waste; ILW: intermediate-level waste; LLW: low-level waste; VLLW: very low-level waste; VSLW: very short-lived waste. For a color version of this figure, see www.iste.co.uk/amiard/radioactive.zip*

1.3.3. *The French classification of radioactive waste*

The details of the French classification are as follows. Radioactive waste is classified according to two criteria: mass activity and physical half-life. The "mass activity" criterion divides waste into four groups: *déchets de très faible activité*, called TFA or very low-level waste (VLLW), *déchets de faible activité*, FA or low-level waste (LLW), *déchets de moyenne activité*, MA or intermediate-level waste (ILW) and *déchets de haute activité*, high-level waste (HLW). The "life" criterion is divided into three classes to distinguish between *déchets à vie courte*, short-lived waste (SLW), *déchets à vie moyenne*, medium-lived waste (MLL) and *déchets à vie longue*, long-lived waste (LLW). The combination of the two criteria makes it possible to classify the waste into 12 categories (Table 1.1) [PNG 10].

	Very short life (VSL) **<100 days**	**Short life (SL)** **≤31 years old**	**Long life (LL)** **More than 31 years old**
Very low activity (VLL) <100 Bq.g^{-1}	VLL-SL Morvilliers	VLL-SL Aube Center	VLL-LL Aube Center
Low activity (LLW) Thousands of Bq.g^{-1}	LLW-VSL Beaumont-Hague, Soulaines	LLW-SL Aube Center	LLW-LL (call for applications)
Average activity (AA) Millions of Bq.g^{-1}	AA-VSL Beaumont-Hague, Soulaines	AA-SL Aube Center	AA-LL Bure?
High activity (HLW) Billions of Bq.g^{-1}	HLW-VSL Not applicable	HLW-SL Bure?	HLW-LL Bure?

Table 1.1. *French classification of radioactive waste and storage sites in operation in France (source: modified from [PNG 10, MTE 18]). For a color version of this table, see www.iste.co.uk/amiard/radioactive.zip*

Radioactive waste management simplifies these subdivisions by grouping certain categories to manage them together. In the end, in France, by combining the four levels of activity with the three ranges of radioactive periods, six categories of waste are distinguished, defined by an order of April 4, 2014. In addition, this decree defines the nature of the information that nuclear activity managers and companies are required to establish, maintain and periodically transmit to ANDRA.

At present, only two categories have well-defined channels: VLA-SL at Morvilliers and LA-SL and AA-SL at Soulaines in the Aube region (and previously in the commune of La Hague, at the *Centre de stockage de la Manche-CSM*, 1969–1994). The other channels are still being studied, as are certain specific wastes such as tritiated waste, mining waste, sealed sources and graphite waste (see Chapter 5).

1.3.3.1. *Activity levels used in France*

Based on their activity levels, nuclear waste can be classified into the following six categories:

– Very short-lived waste (VSL) is managed by allowing it to decay on site and then it is disposed of in conventional channels. It is therefore not sent to a storage facility dedicated to radioactive waste.

– Very low-level waste (VLLW) comes from the operation of nuclear power plants and research centers, from fuel cycle facilities and research centers. The activity level of this waste is generally less than 100 $Bq.g^{-1}$. However, the management of this waste justifies radiation protection monitoring.

– Low-level and intermediate-level short-lived waste (LL/IL-SLW) come from the operation and dismantling of nuclear power plants and research centers and, for a small part, from biomedical research activities. The activity of this waste is between a few hundred $Bq.g^{-1}$ and 1 million $Bq.g^{-1}$.

– Long-lived low-level waste (LL-LLW) consists mainly of graphite waste and radium-bearing waste. Graphite waste has an activity of between 10,000 and 100,000 $Bq.g^{-1}$, essentially long-lived beta emitting radionuclides. It comes from the dismantling of first-generation nuclear power plants (UNGG). Radium-bearing waste, mostly from non-nuclear industrial activities, is mainly composed of long-lived alpha-emitting radionuclides and has an activity of between a few tens of $Bq.g^{-1}$ and a few thousand $Bq.g^{-1}$.

– Long-lived intermediate-level waste (LL-ILW) comes mainly from spent fuel reprocessing activities. It is technological waste (used tools, equipment, etc.), waste from the treatment of effluents such as bituminous sludge and structural waste, the shells and end caps that make up the nuclear fuel cladding, packaged in cemented or compacted waste packages. The activity of this waste is of the order of 1 million to 1 billion $Bq.g^{-1}$.

– High-level waste (HLW) also consists mainly of vitrified waste packages from the reprocessing of spent fuel. These waste packages concentrate the great majority of radionuclides, whether fission products or minor actinides. The activity level of this waste is of the order of several billion $Bq.g^{-1}$ [JOR 14].

1.3.3.2. *French radioactive waste systems*

As Table 1.1 indicates, not all categories of waste have their storage site yet closed in France. We will detail this aspect later (Chapter 5).

Two important aspects condition the classification of radioactive waste. The first aspect is that there is no single classification criterion for determining a waste class. It is indeed necessary to study the activity of the different radionuclides present in the waste to position it in the classification. However, in the absence of a single criterion, the wastes in each category generally fall within a range of mass activity indicated below.

The second aspect is that a particular type of waste may fall into a defined category but not be accepted in the corresponding management channel because of other characteristics (e.g. its chemical composition or physical nature, such as radium-bearing waste that emits a radioactive gas, radon-222). Consequently, the waste category is not necessarily assimilated to its management channel [AMI 13].

1.3.3.3. *Hospital radioactive waste*

With respect to hospital radioactive effluents, French legislation is very strict and requires the intervention of official institutions, in particular ANDRA, for the conditioning, elimination, transport and storage of this waste [FRE 01, ACR 12]. This statement must be moderated, however, in view of the increase in practices involving radionuclides. The next radionuclides to be used will be beta and especially alpha emitters, which have a limited range in living matter. Recently, research is therefore exploring a number of products under development using isotopes such as lutetium-177, promethium-149, bismuth-212, bismuth-213, astatine-211, radium-223 and polonium-210.

1.3.3.4. *Harmfulness of radioactive waste*

For France, the IRSN [IRS 18b] proposes a methodology and possible criteria for assessing the harmfulness of radioactive materials and waste. In order to make the indicators understandable to a wide audience, the situations are defined to respect a minimum degree of realism. Their choice also aims to cover the main exposure routes and a diversity of contexts.

Four situations are considered, the first two of which involve the presence of an individual in a room containing a package of radioactive waste or radioactive material, whether intact or damaged. The last two situations concern the dispersion of the package in the environment and the impact on an entire local human population or the impact on an aquatic ecosystem.

The report also provides an example of the application of the method for three families of waste (vitrified HA, bituminous MAVL and FAVL ^{14}C). The annual impacts after 100 or 1,000 years are provided and proposals are made for broader deployment, making it possible in the long-term to have an indication of the harmfulness of each of the families defined in the national inventory of radioactive materials and waste [IRS 18b].

1.3.4. *American classification*

The American classification of radioactive waste has three classes (A, B and C) based on the maximum activity of a given radionuclide (Table 1.2).

Radionuclide	Class A	Class B	Class C
^{3}H	40	MC	MC
^{14}C	0.8	–	8
^{60}Co	700	MC	MC
^{90}Sr	0.04	150	7,000
^{99}Tc	0.3	–	3
^{129}I	0.008	–	0.08
^{137}Cs	1	44	4,600
All radionuclides with half-life <5 years	700	MC	MC
α emitters with a half-life >5 years	10		100
^{241}Pu	350		3,500
^{242}Cm	2,000		20,000

Table 1.2. *Excerpt from the US NRC classification of radioactive waste based on maximum concentrations of radionuclides and expressed in Ci.m^{-3} (source: [BLA 01]). MC: maximum concentration (no limit for this class)*

1.3.5. *British classification*

The British classification of radioactive waste adopts the IAEA classification into five categories by defining its own criteria for activity levels (Table 1.3).

Waste classes	Characteristics of this class
VLLW, small volume	Waste of 0.1 m^3 that can be disposed of with regular garbage if it contains less than 400 kBq of activity, as well as hospital and university waste. For waste containing carbon-14 and tritium, the activity limit is 4,000 kBq
VLLW, large volume	Radioactive waste with an upper limit of 4 MBq per ton (not including tritium) is disposed of in specified landfills. For waste containing tritium, the upper limit is 40 MBq per ton
LLW	Containing radioactive material other than that suitable for disposal with ordinary waste, but not exceeding 4 GBq per ton of waste or 12 GBq per ton of β and γ activity
ILW	Waste with radioactivity levels above the upper limits for LLW, but which does not generate heat
HLW	Wastes in which the temperature can increase significantly due to their radioactivity, so this factor must be taken into account in the design of storage or disposal facilities

Table 1.3. *The British nuclear waste classification system (source: [OJO 14, RAH 15])*

1.3.6. *Russian classification*

The Russian classification of radioactive waste is based on a division into three classes according to the specific activity of various categories of radionuclides (Table 1.4). The limits of the categories are high.

Category	Specific activity ($Bq.g^{-1}$)			
	Tritium	Beta (except ^3H)	Alpha (except transuranium elements)	Transuranium elements
Low activity	10^6–10^7	$<10^3$	$<10^2$	<10
Average activity	10^7–10^{11}	10^3–10^7	10^2–10^6	10–10^5
High activity	$>10^{11}$	$>10^7$	$>10^6$	$>10^5$

Table 1.4. *Practical classification of radioactive waste in Russia (source: [OJO 14])*

1.3.7. *Comparisons of the various classifications*

Various comparisons can be made between the classifications of radioactive waste used by different countries.

1.3.7.1. *American classification and IAEA recommendation*

The classification recommended by the IAEA and that applied by the United States have no overlap (Table 1.5).

<table>
<tr><td>NRC</td><td colspan="2">Class A</td><td colspan="2">Class B</td><td>Class C</td><td colspan="2">Excess C or GTCC</td></tr>
<tr><td>IAEA</td><td>VLLW</td><td colspan="2">LLW</td><td colspan="3">ILW</td><td>HLW</td></tr>
</table>

Table 1.5. *Comparison of IAEA ([IAE 09a], GSG-1) and NRC ([NRC 15]) classifications (source: [NEA 16a])*

1.3.7.2. *Comparison between the Belgian, French and Canadian radioactive waste classifications*

In Belgium, class A waste has a specific destination and class B and C waste are managed together. In France, the VLLW and LLW-SL categories are managed together, the AA-LL and HALL categories are managed together, while the FA-VL category is managed independently. For the three states, a distinction is made between current waste and historical waste [PAR 18].

	Belgium	**France**	**Canada**
Number of categories	3	5	4
Classification by lifespan and activity level	A (LLW) B (ILW) C (HLW)	TFA (VSLW) FMA-VC (LLW) FA-VL (VLLW) MA-VL (ILW) HA-VL (HLW)	LLW (LLW) ILW (ILW) HLW (HLW + spent fuel) Mining waste
Other more vague categories	NORM, T-NORM Radifer Waste from future sanitation Spent fuel Spent MOX fuel	Waste without a channel Fuel and MOX	

Table 1.6. *Comparison of radioactive waste classifications in Belgium, France and Canada (source: [PAR 18]). In brackets, the equivalences with the IAEA classification from 2009 [IAE 09a]*

1.3.8. *Classification of sealed sources*

For sealed sources, the IAEA [IAE 09a] recommends the classifications reported in Table 1.7.

<table>
<tr><th>Type</th><th>Half-life</th><th>Activity</th><th>Volume</th><th>Examples</th></tr>
<tr><td>VSLW</td><td><100 days</td><td>100 MBq</td><td>Small</td><td>^{90}Y, ^{198}Au (brachytherapy)</td></tr>
<tr><td>VSLW</td><td><100 days</td><td>5 TBq</td><td>Small</td><td>^{192}Ir (brachytherapy)</td></tr>
<tr><td>LLW</td><td><15 years</td><td><10 MBq</td><td>Small</td><td>^{3}H, ^{60}Co, ^{85}Kr</td></tr>
<tr><td>ILW</td><td><15 years</td><td><100 TBq</td><td>Small</td><td>^{60}Co (irradiators)</td></tr>
<tr><td>LLW</td><td><30 years</td><td><1 MBq</td><td>Small</td><td>^{137}Cs (brachytherapy)</td></tr>
<tr><td>ILW</td><td><30 years</td><td><1 PBq</td><td>Small</td><td>^{90}Sr (thickness gauges, thermoelectric generators),
^{137}Cs (irradiators)</td></tr>
<tr><td>ILW</td><td>>30 years</td><td><40 MBq</td><td rowspan="2">Small but with a large number of sources</td><td>Pu, Am, Ra (static eliminators)</td></tr>
<tr><td>ILW</td><td>>30 years</td><td><10 GBq</td><td>^{226}Ra, ^{241}Am (gauges)</td></tr>
</table>

Table 1.7. *Examples of the use of the IAEA classification for disused sealed radioactive sources (source: [IAE 09a])*

1.4. Origins of nuclear waste

Radioactive waste has multiple origins, which can be subdivided into three main sources: waste from the fuel cycle contributing to nuclear electricity (NFC, Nuclear Fuel Cycle), waste from other very varied origins (medicine, research, etc.) and waste resulting from a nuclear accident. Fuel cycle waste differs according to whether it comes from upstream or downstream plants or from nuclear power reactors in operation (Figure 1.2).

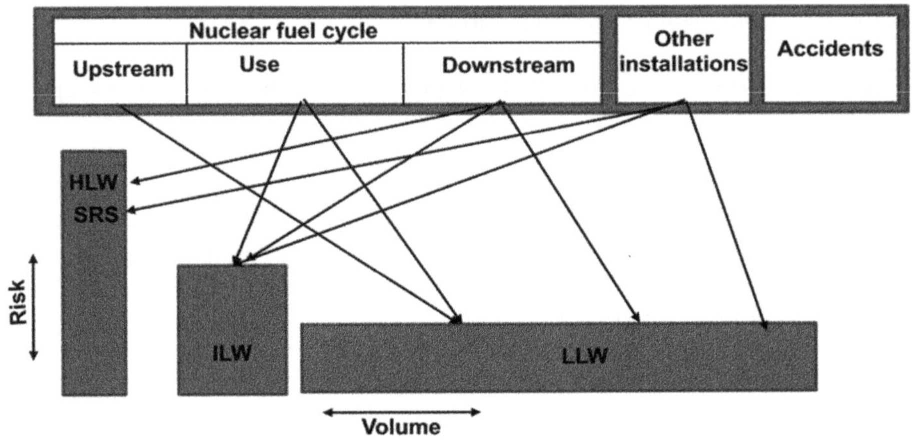

Figure 1.2. *Diagram of the origins of radioactive waste (source: [OJO 14]). HLW: high-level waste; ILW: intermediate-level waste; LLW: low-level waste; NFC: nuclear fuel cycle; SRS: sealed radioactive sources. For a color version of this figure, see www.iste.co.uk/amiard/radioactive.zip*

1.4.1. *The main radionuclides in radioactive waste*

The principal radionuclides in radioactive waste are very varied and can be classified into four categories. These are fission products (H, Se, Br, Kr, Rb, Sr, Y, Mo, Tc, Ru, Rh, Pd, Ag, Cd, In, Sn, Sb, Te, I, Xe, Cs, Ba, La, Ce, Pr, Nd, Pm, Sm, Eu, Gd, Tb and Dy), activation products (C, Cr, Mn, Fe, Co, and Ni) and heavy nuclei (U, Nb and Zr), those that are both fission and activation products (Zr and Nb), heavy nuclei (U, Np, Pu, Am and Cm) and some elements with long-lived radioactive isotopes (C, Zr, Tc, Pd, Sn, I, Cs and Sm) to which are added the five heavy nuclei elements.

1.4.2. *Wastes related to the nuclear fuel cycle*

A distinction should be made between two fuel cycles, the so-called open NFC and the closed NFC, the latter reprocessing spent nuclear fuel in order to reuse the extracted by-products (uranium and plutonium) in other reactors, whereas in the case of the open NFC, the spent fuel is considered as radioactive waste and therefore disposed of. A representation of the two types of fuel cycle is shown in Figure 1.3.

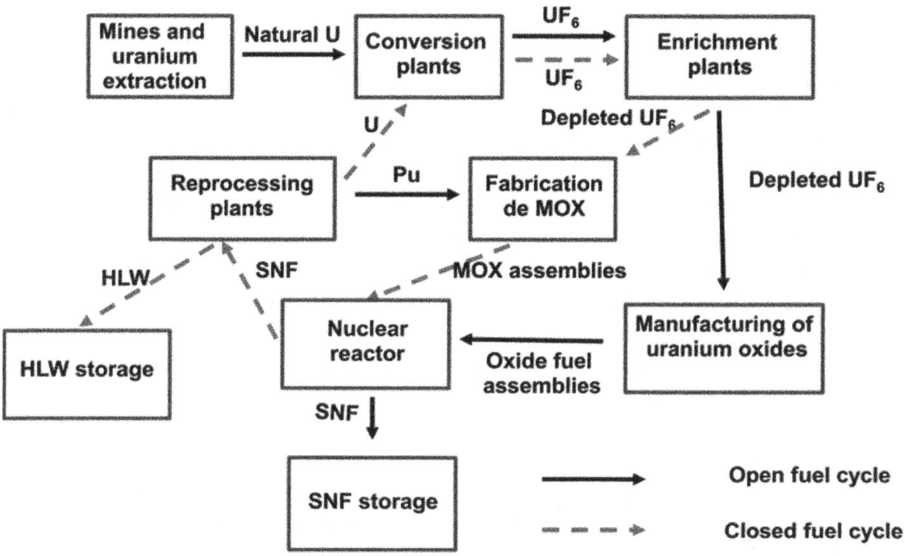

Figure 1.3. *The various stages of the nuclear fuel cycles in open and closed versions (source: [OJO 14]). HLW: high-level waste; MOX: mixed oxide; NFC: nuclear fuel cycle; Pu: plutonium; SNF: spent nuclear fuel; U: uranium; UF6: uranium hexafluoride. For a color version of this figure, see www.iste.co.uk/amiard/radioactive.zip*

The number of states reprocessing civilian spent fuel in 2013 was still six (China, France, India, Japan, the United Kingdom and Russia) with a theoretical annual reprocessing capacity of 5,900 tons to be increased to 6,700 tons [OJO 14]. In 2020, the United Kingdom gave up reprocessing and Japan has had its plants shut down for many years.

The chemical and radioactive composition of HLW varies greatly from state to state. Thus, for transuranium elements, the quantities present in HLW, expressed in $g.L^{-1}$, are 2.0 for the British Magnox reactors, 5.1 for the waste from the La Hague reprocessing plant in France, 7.6 for the WIP (Waste Immobilization Plant) in India, 12.6 for the waste from the Tokai reprocessing plant in Japan and <0.1 for American Hanford waste. Similarly for fission products, the quantities expressed in $g.L^{-1}$ are 87.0 at La Hague, 1.1 at the Indian WIP, 49.0 for the Japanese Tokai plant and <2.5 for the Hanford waste. This can be explained by the characteristics of the reactors and nuclear fuels used, as well as by the cooling methods used and the reprocessing technologies [OJO 14].

1.4.3. *Nuclear waste from electricity production*

About 90% of radioactive waste comes from electricity generation. This waste is of three types. The first category includes waste of various origins (also called type A waste); these are chemical products, work clothes, tools, etc., generally of low radioactivity (18,000 $t.yr^{-1}$ in France). The second group contains technological waste related to the atomic fission process (also called type B waste); this is fairly highly radioactive waste, consisting in particular of metal structures and zircaloy "shells" (an alloy of zirconium and tin, about 1,800 $t.yr^{-1}$ in France). The last group includes waste resulting directly from the fission process of the atom itself (also called type C waste); these are fission products and actinides (approximately 63 $t.yr^{-1}$ and 1.9 $t.yr^{-1}$, respectively, in France), i.e. volumes of 100–240 $m^3.yr^{-1}$. Still for France, each year the nuclear industry produces more than 1,000 tons of spent fuel that is sent to the Orano (previously Areva) plant at La Hague. A portion is processed each year to extract the plutonium (1%) and uranium (95%) and to condition the residue (4%). This is the stage that produces by far the most radioactive waste [AMI 13]. The plutonium is reused in the manufacturing of new fuels (MOX), which are composed of a mixture of plutonium and uranium oxides. There are currently 2,140 tons of irradiated MOX fuel, while 424 tons are loaded into 900 MW reactors [AND 20c].

The quantities of low- and intermediate-level radioactive waste and the tonnage of spent nuclear fuel generated vary widely among the nuclear technologies (Table 1.8.).

	LWR	BWR	PWR	WWER	RBMK	CANDU	Magnox	AGR
LLW and ILW	100	260	130	320	850	80	1,740	400
Spent fuel	25	22	20	28	42	145	240	29

Table 1.8. *Quantities of radioactive waste ($m^3.GW^{-1}$) and spent fuel ($t.GW^{-1}.yr^{-1}$) generated by the various types of nuclear reactors in operation (source: [OJO 14]). LWR: light-water reactor; BWR: boiling water reactor; PWR: pressurized water reactor; WWER: water-water energy reactor; RBMK: Reaktor Bolshoy Moshchnosti Kanalnyi, CANDU: Canadian dioxide uranium; Magnox: magnesium non-oxidizing; AGR: advanced gas-cooled reactor*

1.4.4. *Nuclear waste related to military activities*

Nuclear defense facilities, many of which are currently being dismantled, have generated waste in the past that has not been treated. This old waste, stored in the facilities of the time, will have to be taken back and conditioned. The current maintenance of nuclear weapons also generates waste in small quantities. This waste is managed in the same way as waste from the civilian industry.

The radioactive waste, Soviet and then Russian, dumped in the Kara Sea in 1993–1994 is relatively large (Table 1.9) [NYF 03].

Waste category	Material	Number of objects	Total activity (TBq)
High activity	Reactors with fuel or containers	7	4,700
Intermediate activity	Fuel-less reactors	10	20
Low or intermediate activity	Containers	6,508	580
	Large objects	154	
	Vessels	15	

Table 1.9. *Radioactive waste dumped in the Kara Sea in the Arctic Ocean in 1993-1994 (source: [NYF 03])*

1.4.5. *Wastes related to medical and industrial uses*

Radionuclides have many uses in medicine and biological research. There are about 23 radionuclides that are used as radioactive tracers for various diagnostic purposes. Other radionuclides are present in sealed sources and serve as sources of ionizing radiation for medical, industrial and research applications [AMI 13]. The types of sealed sources are very varied and there are about 52 types of irradiators [IAE 19a].

Various medical and industrial, civilian and military accidents have occurred with abandoned irradiators [AMI 18, AMI 19].

Some industrial activities, such as chemical treatment related to the production of rare earths or the manufacture of phosphate fertilizers, lead to the concentration of natural radioactivity in a residue that becomes radioactive waste. This is particularly the case for the Rhodia plant in La Rochelle (Charente Maritime).

Many radioactive sources in sealed form are used in medicine to treat cancer (brachytherapy). They are used in industry to radiograph welds to test their integrity, to measure the water content of soil and for many other applications. They are also used in research or medicine to establish diagnoses (scintigraphy) or to treat certain cancers (thyroid) as radioactive tracers, in liquid form. The quantity of waste generated is small, but there are more than a thousand users scattered over the French territory [AMI 13].

1.4.6. *Nuclear waste related to the dismantling of nuclear installations*

The CEA monograph [CEA 17] details the various processes for treating materials resulting from dismantling. During the clean-up and dismantling of a nuclear installation, the various treatments generate a wide variety of wastes, organic wastes, graphite wastes, magnesian wastes and very special wastes such as mercury wastes. High-level waste in sludge or powder form and tritiated waste are also produced.

The volumes of solid radioactive waste generated during the decommissioning of the various nuclear fuel cycle facilities are very variable, with a clear preponderance from the deconstruction of nuclear power reactors (Table 1.10).

Step	Type of waste	Quantity ($m^3.GW^{-1}.yr^{-1}$)
UF_6 conversion	Solid	0.5–1
UF_6 enrichment	Solid	5
UO_2 manufacturing	Solid	1–2
Reactor	Solid	375
Reprocessing	Solid	5

Table 1.10. *Quantities of radioactive waste generated during the decommissioning of various nuclear fuel cycle facilities (source: [OJO 14])*

Ojovan and Lee [OJO 14] quantify these various categories of waste from the dismantling of a nuclear power reactor (Table 1.11).

Step	Type of waste	Quantity ($m^3.GW^{-1}.yr^{-1}$)
Miscellaneous (scrap metal)	Solid	15
Sludge	Solid	0.02
Effluents with tritium	Liquids	70
HLW	Liquids	28
ILW	Liquids	25
LLW	Liquids	15
LLW	Solid	65

Table 1.11. *Typical waste during reactor shutdown (source: [OJO 14])*

1.4.7. *Waste from nuclear accidents*

The NEA Expert Group on Fukushima Waste Management and Decommissioning Research and Development (EGFWMD) was established in 2014 to advise Japanese authorities on the management of large quantities of on-site waste with complex properties and to share their experiences with the international community [NEA 16a].

1.5. The global radioactive waste balance

Radioactive waste inventory data are an important element in the development of a national radioactive waste management program because they affect the design and selection of final disposal methods.

The inventory data are generally presented as quantities of radioactive waste in different waste classes, according to the waste classification system developed and adopted by the country or national program in question.

The diversity of classification systems among countries has limited the comparability of waste inventories and made it difficult to interpret waste management practices, both nationally and internationally. To help improve this situation, the Nuclear Energy Agency has developed a methodology that ensures consistency in national radioactive waste and spent fuel inventory data when submitted. This report is a follow-up to the 2016 report [NEA 16b] that presented the methodology and layout for spent fuel submission. It now extends this methodology and layout to all types of radioactive waste and the corresponding management strategies [NEA 17d].

National radioactive waste management programs require very large amounts of data and information across multiple and disparate disciplines. These programs tend to span a period of several decades, resulting in a serious risk of data and information loss, which in turn can threaten the production and maintenance of robust safety records. The NEA has taken the lead in creating a Radioactive Waste Repository Metadata Management (RepMet) project [NEA 18a].

In 2011, Ojovan and Lee [OJO 14] estimated 68.10^6 m^3 of waste stored and $76,10^6$ m^3 of waste disposed (Table 1.12).

At the end of 2013, the quantities of spent fuel discharged from nuclear reactors amounted to 367,600 metric tons, of which about half was stored in wet form, one-third needed to be reprocessed, and the rest was stored in dry form [IAE 18a] (Table 1.13).

Waste category	Stored waste (m^3)	Disposed waste (m^3)
VLLW	153.10^3	113.10^3
LLW	$56{,}663.10^3$	$64{,}792.10^3$
ILW	$8{,}723.10^3$	$10{,}587.10^3$
HLW	$2{,}743.10^3$	72.10^3
Total volume	**$\sim 68.10^6$**	**$\sim 76.10^6$**

Table 1.12. *Global estimate of the global radioactive waste inventory in 2011 (source: [OJO 14])*

Geographical area	Wet storage	Dry storage	Reprocessing	Total
Africa	850	50	NP	900
Eastern Europe	28,600	7,700	3,200	40,000
Western Europe	37,000	4,600	108,000	154,100
Far East	32,100	5,700	8,600	46,400
North America	79,300	41,900	NP	131,200
Latin America	3,000	2,000	NP	5,000
Grand total	180,800	56,900	120,300	367,600

Table 1.13. *Quantities of discharged spent fuel (in ton) at the end of 2013 (source: [IAEA 18a]). NP: not provided*

The storage of spent fuel is carried out for 81% near the producing reactor (59% under water and 22% dry) and for 15% far from this reactor (13% under water and 2% dry), and for the remaining 4% the storage is not known [IAE 18a].

Figure 1.4 highlights the significant quantities of solid radioactive waste worldwide. The less hazardous categories of waste (VLLW and LLW) are larger than the more hazardous ones (ILW and HLW). However, it should be noted that the final solutions are more effective for the former categories compared to the solutions not found for the more hazardous ones.

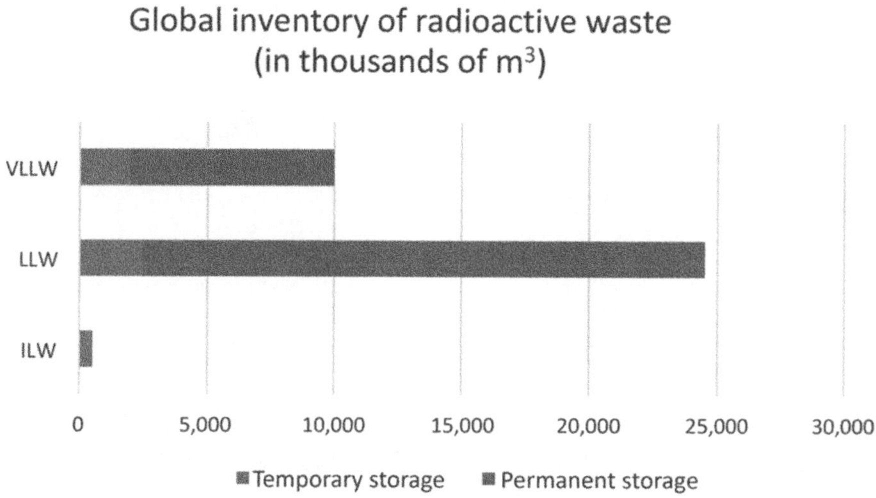

Figure 1.4. *Summary of global inventories of solid radioactive waste in storage and disposal (source: [IAE 18a]). For a color version of this figure, see www.iste.co.uk/amiard/radioactive.zip*

The distribution of solid waste at the end of 2013 by major waste categories and by geographical area is presented in Table 1.14.

Geographical area	VLLW	LLW	ILW	HLW
Africa	7,000	20,000	1,000	0
Eastern Europe	15,000	2,479,000	101,000	7,000
Western Europe	224,000	355,000	269,000	6,000
Far East	5,000	331,000	4,000	0
North America	2,105,000	248,000	84,000	8,000
Latin America	0	37,000	0	0
Middle East and South Asia	0	3,000	0	0
East Asia and Pacific	0	5,000	1,000	0
Grand total	2,356,600	3,479,000	460,000	22,000

Table 1.14. *Quantities of solid waste (in m^3) at the end of 2013 (source: [IAE 18a])*

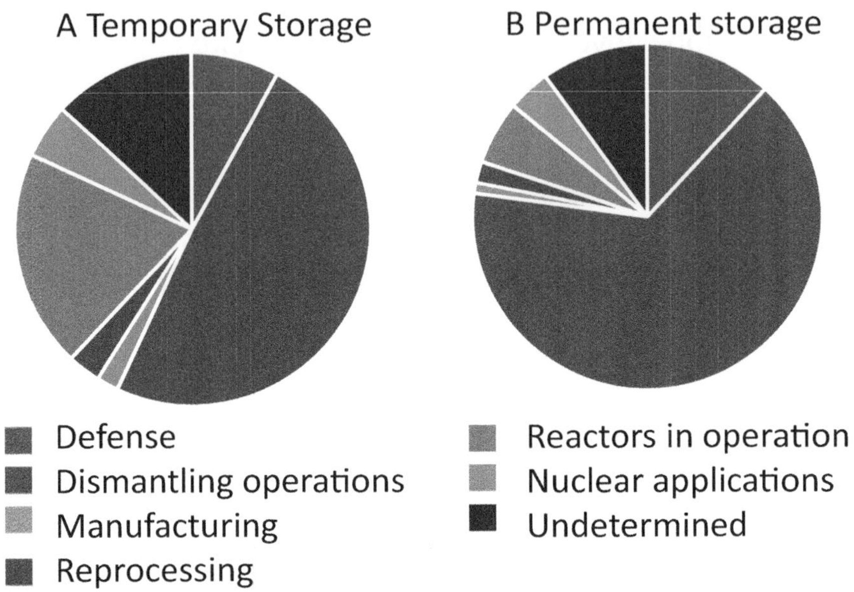

Figure 1.5. *Global origins of radioactive waste in 2013 for A) storage and B) final disposal (source: [IAE 18a]). For a color version of this figure, see www.iste.co.uk/amiard/radioactive.zip*

Worldwide, the majority of radioactive waste comes from dismantling operations (49% and 66%, respectively, depending on whether the storage is interim or final) (Figure 1.5).

Globally, the volumes of radioactive waste at the end of 2013, both solid and liquid, in interim and final storage, for the various categories are shown in Tables 1.15 and 1.16. LLW is the largest category.

Category	Solid	Liquid	Total
VLLW	2,356,000		2,356,000
LLW	3,479,000	53,332,000	56,811,000
ILW	460,000	6,253,000	6,713,000
HLW	22,000	2,786,000	2,808,000

Table 1.15. *Radioactive waste in temporary storage globally at the end of 2013 (in m^3) (source: [IAE 18a])*

The global waste volumes that are permanently stored are 69% for LLW, 29% for VLLW and only 1.63% for ILW and 0.06% for HLW.

The distribution of total activity is 95% for HLW, 3% for ILW, 1.5% for LLW and 0.5% for VLLW [IAE 18a].

Category	Solid	Liquid	Total
VLLW	7,906,000		7,906,000
LLW	20,451,000	39,584,000	60,035,000
ILW	107,000	8,628,000	8,735,000
HLW	0	68,000	68,000

Table 1.16. *Radioactive waste in final storage globally at the end of 2013 (in m^3) (source: [IAE 18a])*

1.6. Conclusions

Radioactive waste is a radioactive substance for which no further use is planned or envisaged. This definition has many implications and divides states into two groups depending on whether their nuclear fuel cycle is open or closed. In the first case, all spent fuel is considered as waste; in the second case, it is radioactive material that can still be used.

The IAEA proposes a classification of radioactive waste based on two criteria: the level of mass activity of the waste and the physical period or half-life of the radionuclide present in the waste. Each state has adopted its own classification, which has a strong influence on the management of radioactive waste.

The origin of radioactive waste is multiple. All activities, whether military or civilian, generate radioactive waste. However, it is the decommissioning and dismantling of nuclear installations that generate the largest volumes. The reprocessing of spent fuel, carried out by a small number of states that have adopted the closure of the nuclear fuel cycle, generates larger volumes of waste, all categories combined, but less high-level waste than the open cycle. Indeed, in the case of an open fuel cycle, the volume of spent fuel is large and this waste is hazardous and difficult to store.

A number of historical nuclear wastes, such as military wastes, mining residues and others, pose difficult conditioning and management problems.

2

Nuclear Waste Disposal Methods

2.1. Introduction. How do we get rid of nuclear waste? What solutions are there for nuclear waste in the future?

Whatever the type of waste, there are few solutions to get rid of it; either it is released into the environment hoping for a strong dilution or it is concentrated and put out of reach for living beings by containment, storage and stockpiling. Often, before this disposal, it is necessary to sort and separate the various pollutants that constitute the waste. In the case of radionuclides, other solutions are possible because of their characteristics. The first alternative is to give the waste time, and short-lived radionuclides will disappear naturally. On the other hand, for long-lived radionuclides, the second alternative likely to eliminate them from radioactive waste is their possible destruction by transmutation [AMI 13]. Unfortunately, this final solution is very rarely applicable because it requires separation and preparation of the target.

Between the two extremes, dilution and final disposal, radioactive waste managers have at their disposal several methods and techniques for neutralizing the danger of radionuclides. However, these are always provisional solutions, such as waste transformation. These include volume reduction, concentration, separation or immobilization in various matrices (cement, bitumen, ceramic, glass). These transformations are followed by storage in pools, dry storage or other storage facilities.

For the most dangerous long-lived and high-level waste (HLW), many solutions have been proposed, some of which were not very safe, such as permanent removal into space or immobilization in polar ice. The safest solutions are burial in deep geological layers that are impermeable to water or deep drilling. Some people propose temporary storage solutions in the hope of future technical innovations, but this expectation has a cost (that of the installation and its maintenance) and from an ethical point of view, this choice amounts to leaving it to future generations to manage our waste.

2.2. Nuclear waste management

There are theoretically many methods for managing radioactive waste. However, not all methods are applicable to all categories of waste. Some methods, such as waiting time, are applicable to very short-lived waste (physical life of less than 100 days), while others, such as dilution, are only feasible for VLLW. The case of long-lived waste (period greater than 31 years), and particularly ILW and HLW, is difficult to resolve and will be dealt with in section 2.3.

The transformation of radioactive waste is generally carried out in three stages, pre-treatment, treatment and conditioning to reduce its harmfulness. However, this schema is not rigorous because some stages are not present or are merged into a single stage. The possible treatments are numerous and depend on the nature of the waste (gaseous, liquid or solid), its activity and its chemical composition. The main treatments are the reduction of the volume of the waste, its immobilization and the separation of radionuclides within the waste.

2.2.1. *Dilutions*

The principle of dilution is simple and is regularly used for all wastes. It consists of injecting the toxic substance into one of the environmental compartments (atmosphere, hydrosphere or lithosphere). If the injection conditions are well controlled, the toxic substance will disperse throughout the volume of the environmental compartment.

Several factors limit this method. These are mainly the respective volumes of the waste and the receptor, the behavior of the waste and its toxicity, the inter-compartment transfers and the interaction of the waste with the biosphere [AMI 13].

This method is limited by the volume of the waste. Thus, a natural waste such as human excrement is well diluted on the Earth's crust as long as the number of individuals is limited in relation to the Earth's surface. Beyond that, there is massive organic pollution.

Another limiting factor is the lack of homogeneous dilution in the physical compartment. Furthermore, the atmosphere is the smallest physical environmental compartment. Waste can reach another physical compartment and even the biological compartment. If it is toxic, it can have a significant impact on flora, fauna and human beings.

Very low-level radioactive waste is routinely released into the atmosphere and hydrosphere during normal operation of various nuclear facilities. Industrial operators have annual air or water discharge authorizations with maximum values defined by the relevant regulatory authorities for each state. Discharges from fuel cycle facilities vary considerably. The most polluting plants, both in terms of airborne discharges and effluent discharges to the sea, are spent fuel reprocessing plants such as Sellafield in the United Kingdom (shut down in 2020) and La Hague in France.

Unintentional direct discharges occur during each nuclear accident, and subsequent discharges are sometimes consequential as for the Chernobyl accident and the Fukushima accidents.

Direct discharges are currently only authorized in the case of radionuclides that are not very toxic and difficult, if not impossible, to trap. This is the case, for example, for tritium released from the La Hague reprocessing plant into the English Channel via a 5 km-long sea outfall, discharging 1.5 km from the shoreline. This discharge takes advantage of the strongest tidal current in Europe, the Raz Blanchard, at certain times of the tide to dilute the radionuclides in the direction of Pas de Calais (Coriolis force). This is also the case for krypton-85, which is released into the atmosphere through a 100 m high chimney and takes advantage of the relatively strong and frequent winds in this region to be diluted in the air [AMI 13].

The use of dilution was also practiced in the early days of civil and military nuclear power. This was the case for the Americans at Hanford into the Columbia River. Similarly, the Soviet and then Russian naval authorities of the Far East fleet discharged radioactive liquids from the reactor tanks of nuclear submarines being dismantled into the Sea of Japan until 1993. The main radionuclides released were cesium-137 and strontium-90. Today, such discharges to the sea are rare and have been banned [AMI 13].

Direct discharges into the lithosphere have also been used historically. This is, for example, the case of the Soviets in two sites (DLRWDF, Deep Liquid Radioactive Waste Disposal Facilities), “Severny” at the Zheleznogorsk site and “Seversky” at Seversk [SAF 18].

2.2.2. *Decontamination*

Decontamination is an operation that consists of removing as much radioactivity as possible from radioactive waste in order to be able to manage it more easily afterward, as it has become "cleaner" or rather "less toxic" for organisms. It is then sometimes possible to lower its danger level and even reuse it in another function. In many cases, the radioactive contamination is located on the surface of the solid waste, and it is possible to recover it by various treatments. The major challenge of radioactive decontamination is to carry out this operation while generating a minimum of waste or secondary effluents [CEA 08].

There are many decontamination methods. The IAEA [IAE 17a] distinguishes between mechanical and chemical methods. Mechanical methods include dabbing, washing, scrubbing, brushing, vacuuming, cleaning with vibration or water and steam jets, and even blasting. Chemical methods involve chemical gels or foams (spraying, foam filling). Surface decontamination can be achieved by electrochemical methods (electro-polishing), ultrasonic cleaning or thermochemical methods. A new plasma decontamination method is promising [GUE 15].

For solid wastes, when possible, surface decontamination is achieved with self-drying "aspirable" gels. This process involves treatment using a dry method, which leads to the production of solid by-products which are easy to condition. The gels developed can be described as concentrated colloidal solutions comprising one or more generally mineral viscosity agents, such as alumina or silica, and an active decontamination agent, for example, an acid, a base, an oxidizing agent, a reducing agent or a mixture of these. The gel is applied to the part to be decontaminated, then dried, brushed and vacuumed to collect the contaminated residue.

Decontamination can also be carried out with aqueous foams. These aqueous decontamination foams are complex two-phase fluids that contain approximately 90% air, a surfactant and one or more chemical reagents to dissolve the contaminant deposit adhering to the wall. The foams are stabilized by viscosity agents or co-surfactants or by particles. Decontamination of solids is also achieved by degreasing with acidic surfactant solutions [CEA 08].

For liquid waste, the challenge of decontamination is to optimize the precipitation of the radionuclides present by using the most selective reagents possible at the lowest possible concentrations. One of the most widely used processes is co-precipitation-settling. This process has been adopted in most liquid effluent treatment plants (LETPs). The most commonly used salts for co-precipitation are barium sulfate, to insolubilize or sorb strontium, and iron and copper hydroxides, to trap alpha emitters and ruthenium. The solid particles are then isolated by decantation, and the resulting radioactive sludge is coated in bitumen (banned in France since 2008) or a cement matrix [CEA 08].

For organic waste, one of the possible treatments is to drastically reduce its volume by incineration. The contamination is then concentrated into a small volume mineral ash that can be easily conditioned. Another way is hydrothermal oxidation by supercritical water.

2.2.3. *Reduction of the volume of radioactive waste*

The reduction of the volume of radioactive waste can be achieved by various treatment processes. The main ones are compaction, evaporation, incineration and fusion [AND 18c]. Ojovan and Lee [OJO 14] and Okoshi and Momma [OKO 15] list the various physical and chemical processes that can reduce the volume of liquid, solid or gaseous nuclear waste.

For solids, the main solutions are compaction, incineration and fusion. Incineration, according to the composition of waste, is carried out by aerobic or anaerobic combustion [OKO 15]. Fusion allows a reduction of the volume of waste, its homogenization and a stabilization of the waste. The methods of fusion are varied – plasma, microwaves, induction at high frequency (conductive crucible method, heating by Joule effect) [OKO 15]. In the case of solids, this reduction can also be obtained by mechanical treatments. The main mechanical treatments use specific tools such as shears, electric nibblers, mechanical saws, orbital knives, abrasive knives (wheels, blades, wires, core drills), milling machines, demolition balls, slicers, concrete breakers, pick hammers, abrasive water jets, rock splitters or even explosives. The choice of tool will depend on the nature of the waste (metals, steel, concrete, etc.) and the environment where the work is to be carried out (air or water) [OJO 14].

For liquid wastes, the volumes of liquid radioactive wastes can be reduced by evaporation, chemical transformation (coagulation, flocculation, separation, electrocoagulation), ion exchange (with organic or inorganic exchangers), reverse osmosis, electrodialysis, ultrafiltration and microfiltration. The reduction factors vary according to the method from a factor of 10 to a factor of 10,000 [OJO 14].

The coagulating agents causing complexation are, for example, calcium carbonate or phosphate and barium sulfate for ^{90}Sr, ferrocyanides for ^{13}7Cs and ferric hydroxide for ^{106}Ru, Pu and Am [OKO 15]. Filtration methods include micro-, ultra- and nano-filtration as well as reverse osmosis. Ion exchange operates with anionic or cationic resins. Their efficiency can be very high (99.9%) with a combination of both types of resins [OKO 15].

For gaseous wastes, the same methods as for liquids are used, but the solutions are mainly filtrations, adsorption on carbon or electrostatic precipitations.

Thus, treatment by evaporation and ultrafiltration of liquid effluents makes it possible to concentrate the activity and direct it to the vitrification workshop. The decontamination of waste contaminated with Pu and other alpha emitters makes it possible to reduce its radiotoxicity and downgrade it to a lower category of waste. For fuel assembly structural waste, replacing cementing by compacting allows the volume of the final residue to be divided by a factor of 3. The incineration of certain wastes by the CEA, EDF or Orano, followed by cementing of the ashes, makes it possible to divide the volumes to be stored by a factor of about 12. Similarly, the constant improvement in techniques and means of measuring the radioactivity actually contained makes it possible to better characterize the raw waste and thus to select the most appropriate and least expensive treatment. Finally, rigorous management of the products entering the installations and sorting of the waste produced at the source allow a notable reduction in the amount of waste to be conditioned [PAT 02].

For organic waste, the main treatments used are incineration or chemical oxidation. Various processes for incinerating organic technological waste exist. The CEA operates the IRIS process (installation for research into the incineration of solids) which allows the treatment of a wide range of organic waste. The organic waste first goes through a pyrolysis stage at a medium temperature (500°C) to eliminate the most corrosive gaseous compounds; then, it is put into a calciner (900°C) supplied with oxygen to complete combustion while concentrating the contamination in the mineral ash. The IDOHL process allows the treatment of organo-halogenated liquids and the incineration is produced by an inductive flare. Similarly, the chemical and thermal oxidations of organic waste can be performed in the supercritical water phase (OHT) [CEA 08].

More recently, the combination of plasma incineration followed by vitrification makes it possible to treat many wastes efficiently. The CEA [CEA 18a] has developed the Shiva process, which makes it possible to incinerate and vitrify radioactive waste in a single step. Plasma heating makes it possible to reach extreme temperatures locally and very quickly. Plasma is of great interest for the treatment of waste, as it allows the decomposition of almost all waste, whether it is solid, liquid, organic or mineral. The principle of the process is based on the continuous feeding of fragmented waste onto a molten glass bath. Once on the surface, the waste materials are subjected to a very high temperature, which produces, on the one hand, a direct combustion phenomenon between the oxygen present and the surface of the organic material, and, on the other hand, pyrolysis for the materials not exposed to oxidants. Moreover, this process allows the incineration of chlorinated and sulfurous waste.

2.2.4. *Radioactive waste immobilizations*

The containment of radioactive waste consists of immobilizing radioactive atoms until they are no longer radioactive or, if this is not possible, until they are almost no longer radioactive. The packaging and its capacity to resist radiation damage naturally depend on the activity of the material to be treated. There are many techniques for immobilizing nuclear waste that constitute the first barrier.

The immobilization of liquid or solid radioactive waste is most often done by coating. The matrices used for coating belong mainly to four main categories, cements, bitumens, glasses and ceramics. Some add coating with polymer resins or plasticization [OKO 15].

2.2.4.1. *Cement*

Cements result from the setting of a mixture of anhydrous cement and water (pure paste), and possibly aggregates of various sizes (grout, mortar, concrete associated with scrap metal or metal fibers). The grouts are used for embedding and the mortars for blocking.

Cements are very varied according to their composition. The main types of cements are tricalcium silicates, dicalcium silicates, tricalcium aluminates, tetracalcium aluminoferrites, limes and calcium sulfates [OJO 14]. However, the most widely used cement is Portland cement, a grinding of clinker, an artificial rock made by heating to 1,450 C a mixture of limestone and clay (80–20%) giving rise to four crystalline phases (silicates, aluminate and aluminoferrite) that in the presence of water form hydrated calcium silicates and other minerals. Various materials (blast furnace slag, natural or calcined pozzolans, siliceous fly ash) can be added to Portland cement [CEA 08].

Due to the large variety of possible cements, studies to develop cementitious coating formulations, adapted to different types of waste, have been initiated reflecting the diversity of waste to be conditioned, as well as their diversity of origin, nature and composition. Moreover, the reactivity of waste in a cementitious medium is highly variable. These interactions are mainly of five types: adsorption, precipitation, co-precipitation, acid–base reactions and redox reactions [CEA 08].

2.2.4.2. *Bitumen*

Immobilization in bitumens uses mainly four varieties: bitumens obtained by direct distillation, by emulsion, by oxidation and by cracking. The first bitumen coatings were carried out in France at the CEA Marcoule center in 1966. Since then, many countries have used this method, especially Russia and Japan.

Bitumenization is mainly used to coat co-precipitation sludge from insolubilization treatments of effluents or from evaporation concentrates from chemical treatment of spent fuel. Bituminous packages are used for LLW, ILW and long-lived waste.

The coating process consists of mixing waste in the form of sludge with bitumen in an extruder. The mixture obtained is dehydrated and poured into a steel drum (of about 220 L) where it cools down. The hot extrusion ensures at the same time the dehydration of the sludge, the homogeneous dispersion of the waste and the immobilization of the radionuclides within the bitumen matrix. The mass incorporation rate of the waste in the bitumen is about 40%. From a chemical point of view, the coated waste is mostly composed of water-insoluble salts (barium sulfate, ferrocyanides, cobalt sulfide) and soluble salts (sodium nitrate, sodium sulfate) [CEA 08].

Bitumen coating presents several risks, a short-term fire risk and long-term risks of corrosion, swelling and overflow of the coating due to radiolysis (caused by radiation) which releases hydrogen in particular. Depending on the activity incorporated and the nature of the radiation, a cask produces between 1 and 10 L of radiolysis gas per year. The gas source term becomes less than a liter after a thousand years, following the decrease of activity. Over a thousand years, the cumulative volume of gas is of the order of 1 m^3 per drum. After saturation of the gases (1% by volume), there is swelling up to 1 cm per year and this causes overflows. Swelling can be prevented by a moderate filling rate of the drums, by limiting the activity or by trapping the radiolysis hydrogen with cobalt sulfide [CEA 08].

In the long-term, bitumen is altered by water. Indeed, although bitumen alone is not very permeable to water and dissolved species, the initial presence of salts favors the absorption of water by diffusion and osmosis. Upon contact with water inside the asphalt, the most soluble salts dissolve locally. The formation of pockets of saline solutions leads to the development of porosity, which facilitates the diffusion of dissolved species towards the external leachate [CEA 08].

2.2.4.3. *Vitrification*

Glass is an excellent containment material, particularly for packaging fission product solutions. This material allows the waste to be converted from a liquid to a solid state, reduces the volume in storage and then in disposal and provides a material that meets the safety requirements for storage and disposal. Glass has good thermal stability, chemical durability and self-irradiation properties. In addition, because of its flexible disordered structure, which allows it to contain many chemical elements, glass is attractive for a mixture of fission products. The radionuclides participate in the structure of the glass, so it is not a simple coating, but a confinement on an atomic scale [CEA 08].

The composition of glass varies greatly from country to country, with more than nine different manufacturing methods. The two main categories of glass used are borosilicate and phosphate [OJO 14]. France has chosen alumino-borosilicate glass as a containment material for solutions of fission products resulting from the treatment of graphite-gas and light-water reactor fuels.

Many countries have, or have had, vitrification programs (France, United States, United Kingdom, Germany, Belgium, Russia, Japan, China, Slovakia, India, South Korea and Italy). The oldest ones date back to 1985. Most are used for HLW, but some states also use them for LLW and ILW [OJO 14].

The industrial vitrification techniques used in the world are mainly of two types. The continuous vitrification process in a metal pot is used in France and the United Kingdom, as well as in India, but in its discontinuous version. The ceramic furnace process in which the glass is heated by an electric current circulating between electrodes (LFCM, Liquid Fed Ceramic Melter) is characterized by the use of a very large ceramic furnace with a longer lifespan than metal pots (up to six years) but this is difficult to change at the end of its life and represents a very large waste. This process has been implemented in facilities now shut down in Belgium (Pamela) and in the United States (West Valley), and is currently in operation in the United States (Savannah), Japan (Tokai Mura) and Russia (Mayak) [CEA 08]. Others were planned in Germany (Karlsruhe) but have since been abandoned, and in Japan (Rokkashô-mura) where it should be put into service in 2022, 24 years late, but without certainty because Japan has unused stocks of plutonium in France and the United Kingdom.

In the United States, vitrification of nuclear waste in glass is currently the preferred waste disposal process. The DOE currently approves only borosilicate (BS) glass for these purposes. However, many nuclear wastes, currently awaiting disposal, have complex and diverse chemical compositions, and often contain components that are poorly soluble or chemically incompatible in BS glass [DAY 04].

In the future, vitrification must evolve because it has several limitations (reduced life and capacity of furnaces, difficult dismantling of vitrification installations). To overcome all these limitations, a new melting technology has been developed. It uses a cooled metal crucible with direct heating in the glass by induction. This technology enables reaching higher melting temperatures (1,200–1,400°C) paving the way to the production of new containment matrices [CEA 08]. This technique, known as the cold crucible, has been used at La Hague to treat dissolved solutions kept until now in tanks, the uranium–molybdenum fuels (vitrified UMo waste).

2.2.4.4. *Ceramics*

Vitrification is now the key solution for conditioning HLW. However, the conditioning of certain radionuclides in a glassy matrix can be difficult for reasons of low solubility in the glassy network or high volatility during high temperature processing. In this case, ceramic matrices can be an interesting alternative to vitrification.

Ceramics are very varied. The most frequently used are based on silicates, phosphates and various oxides, simple or complex [OJO 14].

Ceramic and glass-ceramic matrices are developed for the conditioning of various wastes. Thus, various ceramics are used for minor actinides (zirconolite, britholite, monazite/brabantite, thorium phosphate diphosphate), for cesium, hollandite, and for iodine, vanado-phospho-plumbed apatite with targeted incorporation rates of 10, 5 and 7% by mass, respectively [CEA 08]. It is also possible to prepare glass-ceramics based on calcium silicate apatite and rare earths to immobilize actinides [CAU 09].

Irradiation has effects on the properties of the containment ceramics as soon as the mass contamination in alpha-emitting radionuclides is greater than 5×10^{18} $Bq.g^{-1}$ with swelling and amorphization of the ceramics studied (zirconolite, monazite-brabantite). Thus, the macroscopic swelling of zircolonite is about 6% [CEA 08].

2.2.4.5. *Resins and metal compounds*

In nuclear installations, the purification of water circuits (pool water treatment circuits, primary effluent treatment circuits, waste water treatment circuits) is done by demineralization through ion exchange resins (IER). These resins are replaced regularly as part of preventive maintenance or following unusual pollution. They are generally immobilized in concrete containers (EDF) or metal drums (CEA).

2.2.4.6. *Choice of immobilization matrix*

The choice between immobilization in cements or bitumens will depend on the desired qualities of the final matrix. For example, cements are superior to bitumens for compressive strength, waste loading, thermal stability, radiation durability and biodegradation resistance. Bitumens are generally better than cements for leaching resistance and both are equivalent for gas generation. Chemical comparability varies with the type of waste, with cements not recommended for boric acid and chelating agents, and bitumens for solvents and oils [OJO 14].

Overall, bitumen is a good containment matrix with low radionuclide release rates. However, it is an organic matrix that is therefore flammable and sensitive to radiolysis [CEA 08]. It has no future in France following the decision of the ASN [ASN 08]. Since then, a quadripartite group (CEA, ASN, EDF and Orano) has been trying to find a solution [CEA 18a] and an external group is analyzing their solution [ASN 19e].

2.2.5. *The separation of radionuclides*

The separation of radionuclides consists of isolating a family of radionuclides (e.g. minor actinides) or a single radioactive element (e.g. strontium). Separation can be carried out on many types of radioactive waste (gaseous and liquid effluents, etc.) with the exception of vitrified waste. However, it is an operation that is mainly practiced at the level of reprocessing of spent fuel and therefore in countries with a closed nuclear fuel cycle. In 2020, these were China, France, India and Russia (see Chapter 4). In this case, this separation allows the extraction of spent fuel from reactors, uranium and plutonium being considered as reusable materials, and all other radionuclides considered as radioactive waste. This separation has advantages and disadvantages. Thus, by removing the low-level radioactive uranium, which represents about 95% of the mass of the spent fuel, the volume of waste is considerably reduced. The plutonium (about 1% of the mass) can be recovered by burning it in a reactor to produce energy. The main disadvantage is the danger of the multiple operations required for separation. Indeed, the separation of radioactive atoms is a hot chemical process that is only possible with the use of robotics. In addition, reprocessing operations are the most environmentally damaging stage of the fuel cycle and generate large volumes of radioactive waste.

The separation of americium is carried out by hydro-metallurgical processes. At the CEA, the method most used is PUREX, gradually replaced by SANEX and now especially the EXAm process. Pyrochemical processes can also be used [CEA 15].

In theory, separation can be followed by transmutation, but currently in practice the results are disappointing (see section 3.8). However, separation without transmutation can be interesting. This is the case when a radioactive species gives off a lot of heat but disappears quickly, and if it is packaged separately and stored for a long time, it will disappear. Thus, the waste that will eventually have to be buried will give off less heat and be less active. Separation also allows for the reduction of the volume of long-lived waste [AMI 13].

2.2.6. *Packaging of radioactive waste packages*

There are many conditioning processes. One of the first is to create a barrier between the waste and its environment. The packaging will also facilitate the handling of the packages. For the transport of the packages, specific packaging is generally designed to meet various criteria to resist the risks of accidents (fall, fire, etc.).

2.2.6.1. *The creation of barriers for radioactive waste packages*

These barriers will depend on the class of the waste. For VLLW and LLW, the barriers are often simple metal drums or concrete walls. For ILW and HLW, the containment barriers are more resistant to heat, irradiation flow, corrosion and possible environmental aggression. The barriers are often made of steel or even stainless steel. Some states even manufacture containers made of copper that are highly resistant to corrosion.

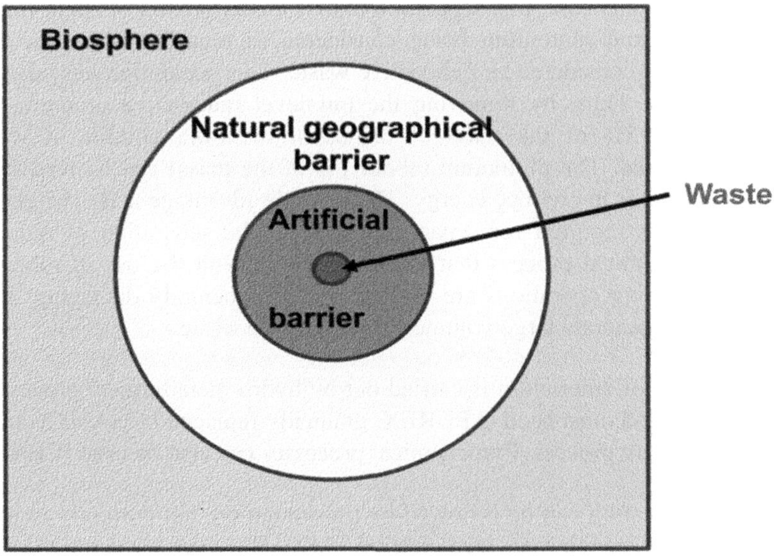

Figure 2.1. *Diagram of two basic barriers of a multi-barrier system in a nuclear waste repository (source: [OJO 14]). EBS: engineered barrier system; NGB: natural geological barrier. For a color version of this figure, see www.iste.co.uk/amiard/ radioactive.zip*

2.2.6.2. *Steel and copper packaging in Sweden*

Metal packaging differs according to the waste category. Metal drums are basic for VLLW and are increasingly sophisticated as the hazardousness of the waste increases. For the most hazardous waste, steel is frequently used. Sweden has developed high purity copper drums that are more resistant to corrosion and that will be used for the storage, in a granite environment, of irradiated fuels only.

2.2.6.3. *Transport packaging*

Depending on the nature of the radioactive waste, the packaging used for transportation must comply with more or less strict rules. The packaging used for the transport of radioactive waste must meet several requirements because the risks associated with transport are not negligible. The level of safety to be achieved is set by the International Atomic Energy Agency's (IAEA) regulations for the transport of radioactive materials. For example, in France, the criteria that packages must meet include resistance to the effects of acceleration and vibration and resistance to a range of temperatures and pressures. In a normal situation, a type A package (medium radioactivity) must be able to withstand a sprinkling, a stacking, a penetration and a drop of 1.2 m. In an accident situation, the type A package must be able to withstand a temperature of 800°C for 30 minutes, an immersion under 15 m of water for 8 hours, a drop of 1 m onto a 150 mm diameter punch and a drop of 8 m. For the most dangerous packages (type B, high radioactivity), the fire resistance test is 60 minutes, the immersion is 200 m, the impact on a rigid target is 324 $km.h^{-1}$ and an embedded position. In addition, the package must withstand a fall of 9 m onto a rigid target, a fall of 3 m onto a cylindrical bar with a conical end and crushing by a 500 kg mass falling onto it from a height of 9 m [IRS 18c].

2.2.7. *Physical decay*

Radioactivity results from an instability of the radioactive atom. This will find a stable form by disintegrating giving risc to a stablc atom. This can bc donc in onc step or in several steps, the child atom being itself radioactive. Radioactive families with a multitude of descendants exist naturally. The last descendant is not necessarily the same chemical element. Thus, in the various families of uranium, the last stable descendants are isotopes of lead.

The decay rate is a physical law that depends on each radionuclide. The physical half-life of a radionuclide, or physical period, corresponds to the half of the atoms of a given radionuclide that decays. Physical half-lives are extremely variable for different radionuclides, ranging from a fraction of a second to millions of years.

2.2.7.1. *The latency method*

This method is very simple and is based on the natural decay of radionuclides. It consists of waiting long enough for the radionuclide concerned to decay to a large extent. Indeed, after 10 physical periods, the initial activity is divided by 1,024. If we can wait that long or longer, the radiological risk due to ionizing radiation will have decreased accordingly. This solution can therefore only be applied to relatively short-lived radionuclides. It is particularly relevant for radionuclides with a physical half-life of less than 100 days, as is often the case for radionuclides used in medicine as radioactive tracers.

2.2.7.2. *Storage*

Storage, or interim storage, of radioactive waste is the operation that consists of placing it temporarily in a facility set up for this purpose in order to allow it to be put on hold, regrouped, monitored or observed. In the case of irradiated fuel, two main types of storage exist: storage under water in pools and dry storage in specific storage spaces.

For low and medium activities, storage is generally on the surface or in the subsurface.

2.2.7.3. *Release thresholds*

Clearance levels are values set by the competent authority or in national legislation, expressed in terms of activity concentration, at or below which materials from practices subject to notification or licensing may be exempted from compliance with the requirements of the legislation. They allow different materials to be "released" and to "leave" the facility where they were produced without special authorization or subsequent control of radioactive materials. The release threshold values are defined by radionuclide and are based on exposure scenarios (internal and external) for workers and the public. The approach used to derive the values is to consider all possible exposure situations, limiting the maximum individual annual dose to any member of the public to 10 $\mu Sv.yr^{-1}$. This value is based on health risk coefficients for the development of cancer. According to the ICRP, this maximum coefficient is $5.7.10^{-2}$ for 1 Sv.

The release thresholds are the subject of international recommendations from the IAEA [IAE 04] and of European rules issued by EURATOM (European Directive 2013/59/Euratom) [UE 14].

Several European countries have enacted their own release thresholds, such as Belgium, Denmark, Finland, Germany, Italy, Latvia, Russia, Spain, Sweden and Switzerland. On the other hand, France, Iceland and Norway do not accept the release thresholds. For France, the Environmental Code prevents the addition of radioactivity in the environment.

There may be significant differences in release limits between states. For example, the thresholds in Japan, when the material is not considered radioactive, are different from the IAEA recommendations [IAE 04]. Thus, the values are, respectively, in Bq.g^{-1} for Japan and the IAEA for ^{3}H 60 and 100, for ^{14}C 4 and 1, for ^{238}Pu, ^{239}Pu and ^{240}Pu, ^{241}Am: 0.6 and 0.1 [TAC 15].

2.2.8. *Final storage*

For all ultimate radioactive waste, a final solution must necessarily be found. The solution will obviously depend on the danger of the waste, as well as the volumes to be stored. Historically, final solutions such as immersions were used. Since then, they have been prohibited.

2.2.8.1. *Immersions*

The IAEA has supported disposal of low- and intermediate-level nuclear wastes by ocean dumping. Although disposal is essentially a dispersal/dilution strategy, rather than a containment strategy, the IAEA has recommended that waste packages be designed to ensure containment of the contents during descent to the seabed and to reduce, to the minimum reasonably practicable, the subsequent release of radionuclides to the marine environment. Since the entry into force of the Convention on the Prevention of Marine Pollution by Dumping of Wastes and Other Matter (known as the London Dumping Convention, LDC) in 1975, these operations have been regulated worldwide. Among other things, the Convention prohibits the dumping of high-level radioactive wastes (HLRW) and requires a special permit for the dumping of LLW. On May 12, 1983, at the seventh consultative meeting of the contracting parties, an amendment to the appendices was proposed, the purpose of which was to definitively prohibit all dumping of radioactive waste at sea [CAL 89].

During the 1950s, some of the waste from European and American nuclear power plants was dumped from ships in the Atlantic and between the Channel Islands and the Cape of La Hague (Fosse des Casquets). Approximately 100,000 tons of radioactive waste were dumped in concrete containers on the ocean floor by a dozen countries, mainly the United Kingdom (76.55%), Switzerland (9.64%), the United States (7.67%), Belgium (4.63%), the USSR (proportion not known) and France (0.77%). The latter stopped its submarine deposits in 1973 [AMI 13].

The geographical areas used for radioactive waste dumping were the Pacific Ocean (Northeast 0.55 PBq and West 0.02 PBq) and the Atlantic Ocean (Northwest 2.94 PBq and Northeast 42.31 PBq) [UNI 10]. This will be developed in Chapter 3.

The containers were expected to remain watertight for about 500 years, the time required to reduce their activity to a level where dispersal in the sea would not be a problem. However, they were subjected to both salt corrosion and the high pressures of the ocean floor.

2.2.8.2. *Final storage centers*

For very low- and low-level radioactive wastes, the creation of final disposal centers does not pose serious problems and many states have found solutions. Generally, these centers remain at ground level or are slightly buried and remain continental. This is the case, for example, in France with the *Centre de stockage de la Manche* (CSM) and the *Centre de stockage de l'Aube*. Two countries, Finland and Sweden, have been innovative by building storage centers for low-level radioactive waste deep under the Baltic Sea (SKUT and SKB).

2.2.9. *Transport of nuclear materials and radioactive waste*

Radioactive materials and wastes change location between the site where they are produced, the site where they are processed and finally the site where they are stored. These transports can be reduced or on the contrary change continent. Worldwide, about 20 million shipments of radioactive materials are transported each year on public roads, railroads and ships. About 95% of these radioactive shipments are not related to nuclear energy but are health or industrial packages [WNA 17a].

The safe transport of radioactive materials and radioactive waste is achieved by a three-level defense in depth: robust package design, reliable operations, and emergency preparedness and response. This third line of defense aims to exclude or reduce the radioactive consequences of events that could not be avoided by the other two lines of defense [SOR 15].

The most dangerous shipments of radioactive packages are those of high-level, long-lived radioactive waste and spent nuclear fuel. Shipments of spent nuclear fuel are particularly numerous. At least 25,400 shipments have been counted worldwide, but the number is probably higher than 44,400 of which 10–17% of this total is made by the United States. The amount of spent fuel shipped worldwide at the end of 2015 was at least 87,000 metric tons of heavy metals (Mt_{ML}) and probably more than 109,000 Mt_{ML} [USD 16]. The case of Japan is typical. For example, from 1969 to 1990, there were over 160 shipments of spent nuclear fuel from Japan to Europe. The first shipment of immobilized HLW from reprocessing to Japan was in 1995 and the 12th and final from France was in 2007. The first from the United Kingdom was in 2010 [WNA 17b].

Cases of radioactive contamination on spent fuel (SNF) and HLRW drums and the vehicles transporting them have occurred more frequently than transport accidents. One of the most serious accidents involved a spent fuel cask in 1971 during transport in the United States. The truck overturned and the cask being transported separated from the trailer [USD 16].

2.3. The special case of long-lived radioactive waste management

For long-lived LLW, ILW and HLW, there are theoretically six main methods for managing it and reducing its harmfulness by keeping it away from all living organisms. These are storage, separation–transmutation, storage in boreholes, storage on the seabed, sending into space and immobilization in polar ice.

In the early 1970s, USAEC initiated a broad compilation of various options for managing HLRW. A voluminous report (BNWL-1900) was edited by Schneider and Platt [SCH 74a; SCH 74b; SCH 74c; SCH 74d], summarized by USAEC [USA 74].

Some projects are unrealistic or unfeasible and were proposed in the 1970s to 1990s, such as sending waste into space, shipping waste to a third world country or burying it in the ocean floor [BAT 90]. The technical difficulties of implementation as well as the evolution of ethical considerations and their legal translation have led to the abandonment of several of the options envisaged historically. This is the case for storage on the seabed, sending into space, and immobilization in polar ice, which are no longer the subject of study and research.

For the majority of states, spent fuel is HLRW. For the four states that reprocess spent fuel, it is not waste, but may become so when reprocessing is abandoned, as was the case for the United States and the United Kingdom. On the other hand, reprocessing residues form two categories of ILRW and HLRW. These are the solid wastes from the spent fuel cladding and the liquid effluents from dissolving the fuel in nitric acid, which contain a large number of fission products and transuranics (neptunium, americium and curium).

2.3.1. *Treatment and packaging*

This subject has been discussed for low- and intermediate-level wastes above (sections 2.4 and 2.6). However, for long-lived radioactive waste, the problem is more difficult to deal with. This is particularly true for spent fuel.

2.3.1.1. *The case of spent fuel*

The first containment of spent fuel rods is their cladding. It is, for example, made of zircaloy for second-generation French reactors. On exit from the reactor, the state of the irradiated cladding is different from its new state. Indeed, during irradiation in the reactor, the microstructure of the cladding and structural materials changes both because of irradiation (modification of the dislocation structure) and because of corrosion (oxidation and formation of hydrides). The passage in the reactor results in a significant hardening of the materials. The oxidation depends on the nature of the alloy and the conditions (temperature, irradiation) undergone. It can be 10 µm thick internally and 100 µm externally with a zircaloy-4 alloy and only 25 µm if the alloy is M5^{TM} [CEA 08].

On exit from the reactor, the spent fuel is very radioactive and gives off a lot of heat; it is necessarily deposited for several years under water in a pool located near the reactor in order to cool down. The duration of cooling depends on the rate of combustion of the uranium pellets (e.g. one year for UNE, three years for MOX).

After this period, the fuel is stored dry or still under water, but generally in a centralized site grouping together the spent fuel from the various nuclear power plants by country. Some states reprocessing spent fuel choose wet storage, others dry storage.

During dry storage or geological disposal, the spent fuel undergoes change. At first, the initial containment is maintained and the system remains closed, exchanging only heat and ionizing radiation with the outside. If the containment is broken by rupture of the fuel cladding or degradation of the container, there is an exchange of material with the outside, and the system is then said to be open.

As long as the system remains closed, radioactive decay causes the formation of helium and new elements with different oxidation states. This results in an increase of the internal pressure, the formation of small helium bubbles and cracks in the cladding.

In case of an incident, the system becomes open in an unsaturated medium. In this case, the fuel may come into contact with air or an oxidizing gas, transforming UO_2 into U_3O_8, a powdery solid with a lower density that can degrade the fuel rod [CEA 08].

Several studies have attempted to model the long-term behavior of spent fuel, and various models have been proposed. The main ones are the behavior model of the spent fuel pellet and the model of radionuclide release from a spent fuel in storage. For the first model, it is mainly the formation of helium that causes concerns within the fuel pellets, but only beyond 10,000 years.

Inevitably, especially through corrosion, the failure of the various casings of the package occurs gradually during deep geological disposal. It is inevitable that water from the environment will come into contact with the nuclear package over several millennia. When water comes into contact with the spent fuel, it solubilizes all the radionuclides present in the voids and on the accessible surfaces of the fuel. This results in an instantaneous release of this labile fraction. The release of radionuclides can also occur with the alteration of the fuel matrix. Indeed, UO_2 is stable in anoxic environments but the alpha emitters cause radiolysis of the water at the fuel/water interface and oxidize U(IV) into U(VI). This leads to the dissolution of uranium and the release of included radionuclides. There can also be a secondary precipitation of minerals [CEA 08].

For long-term storage (LTS) and direct storage, spent fuel containers must be specially designed for these projects.

The concept of the container for storage for 100–300 years is identical for UOX fuel, MOX fuel and vitrified waste. It must fulfil two main functions, containment and retrievability, associated with three secondary functions, the ability to evacuate residual heat from the waste or fuel, the guarantee of subcriticality and the ability to be handled. The other functions (biological protection, physical protection, external cooling) are provided by the repository itself.

The solutions chosen by states vary but are generally multi-barrier. For example, France has chosen a container with two barriers in addition to cladding. The first barrier is a casing for a single fuel assembly and there is a second barrier for several casings. The casing is made of passivatable metal (stainless steel, titanium, copper, tantalum, gold, etc., or various alloys such as Hastelloy, Inconel). This choice for the package casing is motivated by two advantages. Passivation or passivity represents a state of metals or alloys in which their corrosion rate is significantly slowed down by the presence of a natural or artificial passive film, compared to what it would be without this film. Unlike localized corrosion, the evolution of homogeneous corrosion is more easily predictable [CEA 08].

In a storage situation and in the long term, the corrosion of the metal casings and the cladding will be sufficient to bring the radioactive waste into contact with the groundwater. The "labile" fraction (a few percent of the radioactive inventory) will then be released. The countries that have chosen direct storage of spent fuel (Sweden, Finland, United States) are faced with this difficulty. To solve it, all of them rely on sophisticated engineered barriers to ensure the long-term containment of radionuclides. The Swedish concept uses steel–copper containers capable of resisting corrosion for a very long time. But the control of metallic impurities and welds must be very thorough [CEA 08].

2.3.1.2. *The case of acid solutions of fission products and actinides*

If the spent fuel is reprocessed, as is being carried out in China, France, India and Russia, the highly radioactive residues must be immobilized in a matrix that resists damage from irradiation for thousands of years. The first stage involves highly radioactive nitric acid solutions, which are initially stored in silos where they are continuously agitated (to evacuate the hydrogen produced and avoid the accumulation of insolubles on the walls) and cooled. Most of the final solution chosen is the vitrification of these solutions (see section 2.4.3), which makes it possible to reduce the volume of waste and complete its immobilization. France uses special glasses (such as R7T7) as a matrix for HLW. At the end of 2016, there were 5,319 CSD-V containers shipped to clients (98.7% of the total), as well as 15,201 CSD-Vs stored at La Hague [ZER 18a]. In the case of vitrification, the reduction in solubility is good for Al, Si, P, Pb, etc. On the other hand, it is very low for H, He, N, Ne, Ar, Br, Kr, Ru, Rh, Pd, Ag, I, Xe, Pt, Au, Hg and Rn. In addition, some elements such as S, Cl and Mo are not very compatible with glass [OJO 05]. It is also possible to use ceramics, mineral compounds elaborated at high temperatures, which would have the property of being able to incorporate in their crystalline structure atoms having a good chemical compatibility. However, the aging of waste barriers is a subject of uncertainty and concern (see Chapter 7).

2.3.1.3. *The conditioning of metallic structural waste*

Metallic structural waste results from the reprocessing of spent fuel after shearing of the fuel assemblies and consists of pieces of cladding and structural materials (end fittings, grids, guide tubes) that are activated and contaminated by fission products and alpha emitters (Np, Pu, Am, Cm). It is therefore long-lived intermediate-level waste (LL-ILW).

From 1990 to 1995, this type of waste was coated in cement. Since 2002, the La Hague plants have been conditioning this metallic waste by compacting it (volume reduction) to make wafers and then stacking these wafers in stainless steel containers that can go into deep storage.

During the first decades (up to about 300 years), the package remains intact. The release that occurs thereafter can be calculated, radionuclide by radionuclide, as a function of the location of the radionuclide under consideration and the corrosion rate of the material that may contain the radionuclide. Typical containment times are about 10,000 years for zircaloy, 100,000 years for stainless steel and 1,000 years for nickel-based alloys [CEA 08].

2.3.2. *Temporary storage facilities*

Three main categories of HLW must be managed. These are liquid waste, solid or solidified waste and spent fuel assemblies. Liquid waste is generally stored in controlled silos before being solidified, most frequently vitrified. Solid waste is temporarily stored in dry storage facilities with artificial ventilation to limit the temperature to below 510°C.

Spent fuel assemblies from nuclear power reactors must undergo an intermediate storage phase following their discharge from the reactor. Indeed, their initial thermal power is too high and a decrease in the activity of the radionuclides contained in them, allowing a gradual reduction in this power, is necessary in order to be able to transport them first and then to use them in the chosen management method.

After this transitional period, which is necessarily under water in a pool near the reactor, the operator has two possible options. Either it continues to store the spent fuel under water or the spent fuel is placed in a dry storage system.

Four fundamental safety functions must be respected (radiation protection, subcriticality, cooling and containment). Four types of storage are possible, under water or dry, and for each on the site of the nuclear installation or centralized in a single specialized site for several installations [IRS 18a].

For the IRSN, a particularly important point for the safety of spent fuel management operations is the control of aging of zirconium-based fuel cladding, which depends on the temperature reached during storage [IRS 18a].

Several storage concepts have been implemented, above ground, semi-buried or buried, as in Sweden (CLAB). The fuel assemblies are stored either in racks or in baskets.

2.3.2.1. *Underwater storage, local and centralized pools*

In terms of radiation safety functions, radiation protection, handling and storage are relatively easy.

Two containment systems are used. The first system includes the rod cladding and the surrounding water which is purified and cooled, the second system includes the pool hall which is ventilated. The cooling system is provided by the pool water which is filtered and cooled continuously. Evaporation losses must be compensated. Subcriticality is ensured by the geometry of the storage and a neutrophage material (neutron absorber).

The advantages of underwater storage are effective radiological protection. The fuel is kept at a low temperature (pool water maintained between 40 and 50°C), cooling capacities are important, thermal inertia is significant and the fuel can be easily monitored.

The limiting factors are the risk of fuel uncovering, the use of active systems, the complex design of large buildings with constraints on resistance to external aggression, and the difficulty of locating losses in the first barrier (cladding) and leaks in the metal sealing skin (liner) [IRS 18a].

France has chosen temporary storage in pools. This choice is first of all linked to the decision to treat the spent fuel to recycle the plutonium and uranium. Underwater reprocessing is easier. Because of the plutonium content and its isotopic composition, spent MOX fuel has a higher thermal power and must necessarily be cooled longer.

The major safety requirements for storage in a pool are the maintenance of a sufficient water inventory in the pool and the availability of cooling systems under all plausible circumstances [IRS 18a].

The decisive parameter for waste storage is its thermal power. For example, in France, the transport of nuclear material is forbidden if the thermal power is higher than 6 kW, and dry storage is only authorized if the nuclear material has a thermal power lower than 2 kW. The thermal power depends on the spent fuel, in particular on its initial fissile content, on the burnup rate in $MWj.t^{-1}$, and on the cooling time it has undergone since its unloading [IRS 19a].

2.3.2.2. *Dry storage*

Three main dry fuel storage concepts have been developed worldwide. These are storage in caverns or wells, storage in silos and storage in packages. They can be realized on the reactor site or centralized.

In dry storage in caverns or wells, also known as cells, the structures are made of reinforced concrete and the containers of metal. The vertical loading concept is the MACSTOR (Modular Air-Cooled STORage) from AECL (Atomic Energy Canada Limited). The blocks are in the open air and receive steel containers loaded with CANDU (CANada Deuterium Uranium) bundles. Another concept with vertical loading but with semi-buried silos has been developed by HOLTEC International (HI STORM 100U and UMAX) [IRS 18c].

Dry storage in silos involves concrete structures with cavities in which the metal containers containing the spent fuel are placed. They can be monolithic or modular. One of the characteristic systems of storage in silos with horizontal loading is that developed by Orano TN, the NUHOMS.

Dry storage in containers involves the use of the latter for both transport and storage. They are mainly made of metal or concrete. This type of dual-use packaging has been the subject of numerous concepts since the 1990s (CASTOR and CONSTOR from GNS, TN 24 from TN International, HI-STAR 100 from HOLTEC International and several options from NAC-STC in the United States among others). In Switzerland and Belgium, spent fuel is stored in metal containers of the TN 24 family from Orano TN, such as the TN 24 DH in the Doel nuclear power plant in Belgium.

Centralized dry storage has been the subject of many concepts, above ground, semi-buried or buried with wells, containers or silos. Radiation protection safety functions are provided by concrete cells, the closure plugs of the shafts and the body of the casks. Containment is provided by the liner and containers or shafts, as well as by drying, inert gases and leak monitoring. Cooling is by natural circulation of outside air. Subcriticality is ensured by the storage geometry and a neutrophage material.

The advantages are the radiological protection provided by the structures, passive cooling which is suitable for very cold fuels and very simple operation. In addition, in the event of an accident, the number of fuels involved is smaller, the thermal power is lower and the consequences for the environment are more limited.

There are three main limiting factors. They are the low capacity for evacuation of the thermal power of the spent fuel (about 2 kW per assembly for MOX, which needs to be cooled for several decades in a pool), the high temperature of the fuel (350–450°C) and the difficulty of monitoring the fuel [IRS 18a].

Dry storage has been carried out in the United States since 1986 and is highly developed there. Spent fuel is currently stored dry in 34 states at more than 60 generally licensed sites and 15 sites with specific licenses. The NRC has developed container licensing requirements through a public comment process to provide a sound basis for ensuring protection of public health and safety and the environment [NRC 16].

Orano TN in the United States is developing several types of containers for both transportation and dry storage. Its NUHOMS® EOS 89BTH DSC (EOS, Extended Optimized Storage; DSC, Dry Shielded Canister) model accommodates spent fuel assemblies from BWRs. The NUHOMS® EOS 37PTH DSC holds 37 PWR spent fuel assemblies. Orano TN's NUHOMS® MATRIX (HSM-MX) offers customers an optimized two-tier horizontal storage system with a proven loading system that improves the stability of horizontal transfer of spent nuclear fuel [ORA 18a].

2.3.2.3. *Comparison of wet and dry storage*

The two types of storage do not meet the same need. Storage in a pool is imperative for low-cold fuels, while dry storage is well suited for very cold fuels.

The type of spent fuel (UNE, MOX, URE) influences the choice of storage type (with mandatory storage under water at the beginning for sufficient cooling). For safety, the decisive parameter is the thermal power. Storage under water requires more important safety provisions [IRS 18a].

2.3.2.4. *Silos*

The fission product solutions, preconcentrated to reduce their volume, are temporarily stored in stainless steel tanks that are constantly agitated and cooled. Their activity, linked to the treated fuel's rate of combustion, can reach $3.7.10^{13}$ $Bq.L^{-1}$ and the power released is significant (up to 7 $W.L^{-1}$). These solutions of nitric acid (1–2 N) are characterized by a high physico-chemical complexity [CEA 08].

The main silos are old and near reprocessing plants such as in Hanford, Mayak, Sellafield or La Hague. In the last site, they disappeared at the end of 2017. Several serious nuclear accidents are due to this type of storage [AMI 18, AMI 19].

2.3.3. *Long-term storage*

The concept of long-term storage is the technical response proposed to the political question of reversibility and to the persistent tension between the technically acceptable solution and its rejection by society. It is a question of proposing a choice of safe conservation of nuclear waste over periods of the order of a century, with the possibility at any time of taking back the waste. This recovery is imperative at the end of the maximum duration for which the installation was designed [CAV 02].

Long-term storage of high-level and long-lived nuclear waste must ensure the loading, holding and preservation of packages of this waste over a period of 50–300 years, as well as their retrieval during this period. These functions must, of course, be carried out under regulated safety conditions and with reasonable economic viability. The technical and political objective is therefore to offer future generations, over the next three centuries, a choice of decisions as wide as ours, under safety conditions at least as good as today. This objective is expressed in terms of the safety of operation of an HLW storage facility by a dozen guiding principles set out by Cavedon [CAV 02].

LTS awaits solution. In case of degradation (abandoned surveillance, lack of maintenance), the consequences for the environment and humans would be very serious because, contrary to geological storage, the geological barrier is absent and the arrival in the biosphere will not be slowed down or mitigated.

In long-term storage (LTS), the duration generally used is one to three centuries. However, it is by no means definitive and the packages must be retrieved. Two options are possible: the extension of the life of the current repositories, but this is limited to 100 years, or the construction of new repositories specially designed to obtain a life of 300 years. Studies for secular lifetimes have been carried out at the

CEA (such as the CECER at Marcoule), the limiting factors being the durability of the concrete and the phenomena of metal corrosion [BON 11].

In order to take into account their long life, it is necessary to verify not only the durability of the packages and the warehouses but also the societal risks. These risks are numerous and diverse, from the fall of an airplane to the loss of technical control of future companies. A long-term warehouse must always remain under the company's control. It is therefore necessary to foresee periods when the installation is abandoned without altering either its functions or the characteristics of the waste (mainly vitrified waste). The nuclear wastes relevant to LTS are long-lived intermediate-level waste (LL-ILW), high-level waste (HLW) and spent fuel. The design of the LTS must meet several criteria: durability, robustness to environmental and societal hazards, the components must be as simple and passive as possible, and the installation must have a high degree of inertia.

The installation could be on the surface or in the subsurface in hard rock above the water table. The primary HLW packages differ from the HL packages in their contents, in the cladding matrices (concrete, steel, etc.), in their dimensions and also in the fact that they give off much less heat. On the other hand, HL packages are generally vitrified and give off a lot of heat, especially in the first few years. Spent fuel (UOX, URE and especially MOX) gives off even more heat, especially after three centuries when the differential will be 9.6.

The LTS design must be static and containment must be provided solely by the package itself. Hence, the packaging of several containers that become integral, also creating a second containment barrier. The installation must protect the packages from external aggression, evacuate thermal power and gases, and allow the packages to be retrieved after one or three centuries.

The packages are made of steel or cast iron (Germany) for HL waste and of ceramic or concrete for HL waste. For spent fuel and HL waste, storage is dry, where the packages are placed horizontally in caverns and cooling is entirely passive, being ensured by natural convection through a chimney [IRS 18a]. Corrosion is limited to 350 μm in 300 years. The installation can be in the subsurface with horizontal galleries giving access to vertical shafts containing the packages. Here again, cooling is by natural convection through two galleries below and above the packages. For LL-ILW packages, the installation is on the surface [SIL 06].

2.3.4. *Storage in the seabed*

In addition to disposal of low- and intermediate-level wastes, the IAEA and the NEA have considered, since the 1970s, the possibility of removing high-level and long-lived radioactive waste from any life source by storing it on the seabed. It is possible to distinguish three main methods of storage in the seabed: storage in abyssal trenches, deposition in marine subduction zones and drilling of the seabed.

2.3.4.1. *Storage in abyssal trenches*

The deep sea, especially abyssal trenches, has generally been considered as a remote, sparsely populated and biologically inactive environment, quite suitable for receiving the noxious products resulting from nuclear fission. Better knowledge, however, has highlighted the existence of vertical migrations linking deep benthic communities to those of the upper ocean layers [RIC 78].

The storage of radioactive waste on the seabed consists of keeping the waste away from any human presence by placing it on the ocean floor. Two options have been considered for this. The first is to deposit the waste in areas where the water depth is significant or where sedimentation is rapid. These zones correspond in particular to the abyssal plains of the deep sea. The depth of the water column can reach approximately 5,000–6,000 m. According to this option, the waste would either be placed on the ocean floor until it is covered by sedimentation, or buried in the unconsolidated sediments that cover the bedrock, generally made up of basaltic formations.

The second option is to deposit the waste in the so-called subduction zones where the oceanic plate sinks into the Earth's mantle. This configuration has led to the possibility of sending the waste to the mantle by placing it on the "conveyor belt" formed by the oceanic plates.

The oceans cover more than 70% of the Earth's surface, and while they contain many valuable resources, they also cover, as was thought in the 1970s, some of the most inaccessible and unproductive areas of the planet. These ocean areas could potentially provide convenient, non-penetrative areas for the disposal of HLRW. Bishop and Hollister [BIS 74] thought that the geological stability and relative uselessness of some ocean basin bottoms could be suitable areas for ultimate nuclear waste repositories.

Similarly, fine-grained clay formations in some deep sea regions, far from the lithospheric plate boundaries and productive surface waters, have properties that could be used to permanently isolate radioactive waste. The most important characteristics of these clays are their vertical and lateral uniformity, low permeability, very high cation holding capacity and potential for self-healing when disturbed. The most attractive abyssal clay formation (red oxidized clay) covers nearly 30% of the seafloor and thus 20% of the Earth's surface [HOL 81].

The first studies began in the United States in 1974 [USD 17], and the work then focused mainly on the abyssal plains of the Pacific Ocean. In the context of the time, the idea of storage on the deep sea floor was based on the low interest perceived for these areas from the point of view of human activities. In addition, the conditions considered favorable *a priori* from the point of view of dilution/dispersion and corrosion were guaranteed by the slow marine currents and the low temperatures expected on the ocean floor. Following this preliminary research on underwater burial of nuclear waste, further research showed that disposal in the seabed would work well, and the United States could probably engage in deep burial by drilling without violating applicable international and national laws [BAL 14].

The molecular diffusion coefficients are very low in this medium. They are 0.010 m^2 per year for ^{239}Pu, ^{137}Cs and ^{99}Tc and 0.018 m^2 per year for ^{129}I [HIC 80].

The subseabed disposal project (DSP) was part of an international program that studied the disposal of HLRW in deep ocean sediments [KLE 97]. The studies show that the attenuation (reduction) factors on the maximum individual dose are significant (Table 2.1).

Components	Scenarios			
	Complete system	Defective containers with waste	On the ocean floor	Deposit only (excluding the ocean)
Containers	$2.9.10^2$	-	$2.9.10^2$	$2.9.10^2$
Waste	$2.6.10^1$		$2.6.10^1$	$2.6.10^1$
Sediment	$1.6.10^5$	$1.2.10^9$	-	$1.6.10^5$
Oceans	$1.3.10^{11}$	$1.3.10^{11}$	$5.5.10^{12}$	-
Total	$1.6.10^{20}$	$1.6.10^{20}$	$4.3.10^{16}$	$1.3.10^9$

Table 2.1. *Attenuation factors for high-level waste deposited on the seafloor based on the maximum individual dose standard. The system attenuation required for 10 deposits is $8.5.10^{15}$ (Nares Abyssal Plain, 5,000 year old containers, 2,500 year leachate, 50 m burial depth) (source: [KLE 97])*

The NEA Seafloor Working Group concludes that underwater burial appears to be a technically feasible method of disposal of HLRW or spent fuel. Two options, placement of penetrators (steel tubing shaped like a torpedo) and placement of borings, have been shown to be technically capable of placing waste containers at the required depths in the sediment. Preliminary cost estimates suggest that HLW or spent fuel could be disposed of economically by seabed disposal [OEC 88]. These projects have been abandoned.

The MPG (Mid Plate/Mid Gyre) program demonstrates that the water column is not stable and that there are exchanges between the surface and the bottom. Hence, the research that focused on the possibility of burying waste in marine sediments. The hypothesis was that the sediment was an effective barrier for trapping radionuclides.

Studies on the burial of waste in sediments began in 1976 with the "Subseabed Disposal Program". This international cooperative program under the auspices of the NEA involved the United States, France, the United Kingdom and Japan. Canada, the Federal Republic of Germany, the Netherlands, Belgium and the Organization of European Communities also contributed. The work resulted in two main options: insertion of the waste into the sediment using penetrators and the construction of boreholes, in which piles of conditioned waste would be placed. A depth of 800 meters below the seafloor was targeted, with the top of the container stack located 300 m below the seafloor. The packages would isolate the waste for 500–1,000 years after placement. Long-term containment, for tens of millennia, would be provided by the sediment barrier.

In 1986, the American DOE, the largest contributor to the Subseabed Disposal Program, decided to stop its financing in order to devote itself fully to the "geological disposal" project, thus putting an end to the program [NAD 96].

In 1988, the NEA/OECD working group on burial concluded that the introduction of HLRW into the seabed sediments was technically feasible, adding, however, that before implementing this option, the evaluation of the long-term safety of these operations required further study in order to reduce the uncertainties [CAL 89].

The evolution of the law of the sea and the international initiatives undertaken to protect the oceans have progressively led to a slowing down of research on the storage of waste under the seabed, and then to its cessation following the moratorium on the dumping of waste at sea in 1983 [IRS 19c].

The scientific, technical, legal and institutional feasibility of underwater storage (SSD, Sub-Seabed Disposal) of HLRW was studied from 1976 to 1987 and resulted in a negative opinion. Since 1990, given the resurgence of the problem of nuclear energy in the face of a massive problem of carbon dioxide accumulation and its consequences on climate change, some researchers believe that the subject deserves to be studied again [MIL 14].

Research on abyssal sediments as containment barriers against radionuclide migration continued. The expected application was the storage of HLW produced by the nuclear industry. Simple models of americium migration show that, under nominal storage conditions, the sedimentary barrier is totally effective for americium [BOU 86].

The study of the solid partitioning of plutonium, scrupulously validated, could lead to a profound revision of the reactivity of this element during the diagenesis of anoxic marine sediments [BOU 97, LUC 04, GOU 04]. Plutonium is 75% bound to reactive sulfides and carbonates (as surface complexes or embedded in the carbonate matrix) and it is necessary to propose more realistic processes that better account for the fluxes observed towards the water column (recycling of sulfides, complexation by organic matter during degradation, bioturbation, bio-irrigation [GOU 04]. A significant proportion of the plutonium is loosely bound to sites that exchange readily with seawater, oxidize on contact with hydrogen peroxide (reactive sulfides) or dissolve readily when pH changes. These reactive sulfides are likely to be a source of plutonium for the overlying waters if they are brought close to the interface by bioturbation or in contact with oxygenated seawater by burrowing activity [GOU 05].

Data from the Ravenglass Estuary in the northeastern Irish Sea show that plutonium is primarily bound to geochemical phases targeted by acid-soluble and exchangeable extractants, indicating that a significant proportion of plutonium in these and similar sediments is associated with relatively mobile geochemical phases [LUC 04].

Most of the plutonium activity dissolved in seawater (98% in 2002) is due to its emission to the water column from the sediments. Paradoxically, the work of McDonald *et al.* [MCD 01] showed that most of the plutonium was bound to low reactive phases during diagenesis.

2.3.4.2. *Disposal in the subduction zones of the sea floor*

The concept of continental drift developed in the early 20th century by Alfred Wegener was not accepted by the scientific community until the 1960s with the development of the geological concept of plate tectonics. The earth has several subduction zones where a tectonic plate descends adjacent to the non-descending crust into the central region of the earth.

The idea proposed in the 1970s in the United States was of placing radioactive waste in subduction trenches and thus using a natural mechanism to send it into the Earth's mantle [BOS 70]. This idea was a direct result of this evolution of scientific knowledge. However, in 1972, the geologist E.A. Silver of the US Geological Survey underlined in the magazine *Nature* [SIL 72] the extreme slowness of the subduction phenomenon and the even slower speed of the sedimentation processes. He added that the subduction zones show signs of strong seismic activity (deformations and faults). Finally, in the slow subduction zones, the observations seemed to indicate that the sediments on top of the subducting plate did not penetrate the mantle but were eroded during the plunge of the plate and accumulated at the surface.

However, several US patents have been successively proposed to improve the radioactive waste disposal process [KRU 79, BAI 91, COL 93]. The inventions relate to a method and device for the disposal of nuclear and other toxic wastes involving the placement of the wastes in holes drilled in the tectonic plate at the edge of the subduction zone and allowing the descending tectonic plate to transport the wastes into the center of the earth. Millions of years are required before the waste can be returned to the Earth's environment and, therefore, decomposition or melting of the waste during this time will render it harmless. The device includes re-entry towers that are positioned over the boreholes to load the waste into them in the form of elongated cylinders or in the form of spheres.

Our current knowledge has progressed since the 1970s. Thus, the process of subduction is slow, the plate sinking about 1–10 cm per year (maximum speed in the western Pacific Ocean). The friction produces earthquakes, and the melting of the oceanic crust from a certain depth leads to an ascent of magma and therefore explosive volcanism on the surface (volcanic arc). When the oceanic crust undergoes subduction, its upper part, made up of unconsolidated marine sediments, accumulates on the surface to form a mountain chain (accretionary prism).

The possibility that the waste may remain on the surface with sediments has not been fully clarified to date [LAL 94]. Finally, the conclusions of Alden, in 2020 [ALD 20], raise the difficulty caused by the slowness of the phenomenon of subduction, already mentioned by E.A. Silver [SIL 72]. He points out that, in the Peru–Chile subduction zone, the Nazca plate plunges under South America at an angle of 30° and at about 7–8 cm per year, which makes it the fastest subduction zone in the world. Under these conditions, however favorable, he deduced that it would take 10,000 years for waste to move horizontally 600–700 m and vertically 350–400 m.

The underwater solution to the problem of radioactive waste is strongly opposed by the US federal government, the nuclear industry and environmental interests [NAD 96].

Similarly, this solution was abandoned in the United Kingdom (CoRWM, Committee on Radioactive Waste Management) in 2004 and in Canada [IRS 19c].

Recently, the disposal of high activity materials in subduction zones has been vigorously refuted [BAI 05].

2.3.5. *Geological storage in a deep continental repository*

Since oceanic storage poses several problems, the authorities have turned to deep geological storage but at the level of the continental crust.

The NEA conducted a broad reflection on the issues of deep geologic disposal. Themes addressed by the experts included duties to future generations, regulatory timelines, staged decision-making, optimization roles and best available techniques (BATs), multiple lines of reasoning, safety and performance indicators, recognition of uncertainties and the importance of stakeholder interactions [NEA 10].

Many states have plans for burial in four types of geologic formations. The advantages and disadvantages regarding rock characteristics and the major radionuclide transport mechanisms in these media are presented in Table 2.3.

Geological formation	Rock characteristics	Radionuclide transport mechanisms	Country
Granite, gneiss	Fracturing and groundwater flow in fractures	Advection and some diffusion	Canada, China, Finland, Russia, Sweden, United Kingdom
Salt dome	No fractures or groundwater	No transport	Germany, United States (WIPP)
Volcanic tuffs and lava	Fractures and pores Unsaturated	Water percolation	United States (YMP), United Kingdom (Longbands Farm)
Clays	No fractures Stagnant water in the pores	Broadcasting	Belgium, China, Hungary, France, Russia, Switzerland

Table 2.2. *Deep geologic formations for radioactive waste management (source: [OJO 14])*

The main migration of radionuclides follows that of water and will follow essentially the faults in the rock [AHN 15].

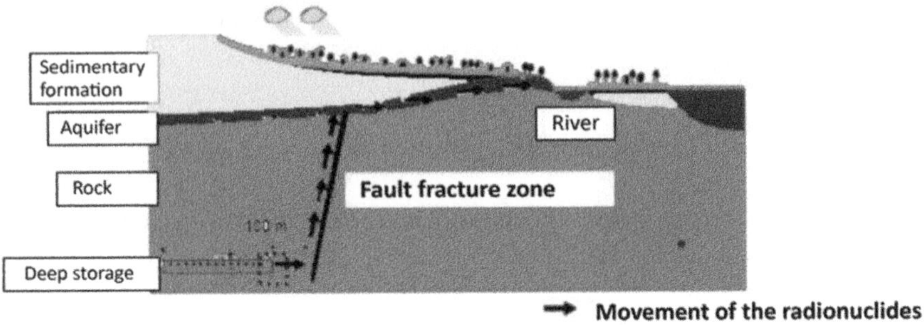

Figure 2.2. *Model of a subsurface migration scenario across a fault fracture (source: [AHN 15]). For a color version of this figure, see www.iste.co.uk/amiard/radioactive.zip*

2.3.5.1. *Final geological repositories*

The concept of geological disposal is simple: it is a matter of placing the waste in a place where its radiotoxicity will no longer be able to act on the environment, and this over a period long enough for the impact on the biosphere to be negligible. But the time to be taken into account is of the order of 10^5–10^6 years. Indeed, ^{239}Pu has a physical half-life of 24,000 years. ^{237}Np is much less abundant but has a physical half-life of 2.10^6 years and ^{129}I has a half-life of 16.1 million years. So only deeply buried sites in a stable and impermeable environment are possible [CAS 02].

The main aim of storage is to oppose the migration of radionuclides with several successive barriers. The first barrier is the containment matrix (glass, ceramic, bitumen, concrete) in which the radionuclides are included. The second barrier is the waste container itself, which lasts for about 1000 years. The third barrier is the engineered barrier used to seal the shafts and access galleries dug during construction of the site and to cover the radioactive packages. The fourth barrier is the geological environment where the site is located, whose role is not only to slow down the migration of radionuclides into the biosphere, but also, if there are leaks, to ensure their dilution in a large volume. The choice of this medium is important because of the role it can play in the chemical stability of the site, which will protect the packages and limit their corrosion [CAS 02]. The nature of the rocks considered is generally salt, granite or clay. The second barrier or engineered barrier is itself often made up of several barriers with the container envelope and the seals. This barrier first includes a liner around the container made of highly absorbent clay, such as bentonite, which has been previously dried and compacted, and in the form of bricks or cylindrical rings.

Hardened clay formations, by the nature and the arrangement of their minerals as well as by their strong compactness, present for the most part the qualities required to ensure an effective confinement of radionuclides. Furthermore, their very low water content (less than 10% by mass) and the very small size of their pores (a few nanometers) make them a very low permeability rock where water circulation, in the absence of fracturing, is very low. Under these conditions, the dominant transport phenomenon of radionuclides is limited to the slow process of molecular diffusion. Moreover, this type of rock is made up of clay minerals carrying negative charges which have the property of being able to strongly retain positively charged radionuclides on their surfaces. On the other hand, all the particular properties of these formations that make them potential host sites make their characterization difficult [AMI 13].

More than granites or salts, clay formations are the most studied at present in Western Europe for possible storage of radioactive waste. A large number of these studies were and still are interested in the migration of solutes through the unaltered rock or through the rock damaged by the excavation of galleries or drillings. Indeed, one of the indicators used to evaluate the safety of an underground storage site is based on the evaluation of the migration times of radionuclides through the rock and into the biosphere, and thus on the knowledge of the transport phenomena within the geological environment.

Most often, only direct flows (diffusion, advection, thermal conduction) are considered and coupled flows are generally ignored. However, phenomena such as thermo-osmosis and/or chemical osmosis are suspected to be at the origin of the hydraulic overloads recorded in Bure clays [AMI 13].

One of the goals of the studies is to evaluate the importance of advection versus diffusion in clay formations. The use of natural tracers has many advantages. In particular, the profiles of natural tracers can be considered as the result of a natural experiment, allowing the characterization of transfers on the spatial scale of the formation and over geological time. A natural tracer is any chemical or isotopic species that is mobile, i.e. whose concentration is modified only by mixing, which excludes all species that adsorb on the surface of minerals, that are not constrained by a thermodynamic equilibrium reaction with a mineral phase and those whose isotopic ratios fractionate in the liquid phase [BEN 10, BEN 13].

Deep geological storage presents anthropogenic, geological and climatic risks. The four main anthropogenic risks are the will to harm (terrorism), the attempt to recover the stored materials, accidental access linked to the presence of exploitable natural resources and purely accidental access (phreatic drilling). The geological and climatic risks are mainly meteorite impact, volcanic eruptions, the appearance of a major fault, intense and abnormal seismicity, erosion by glaciers and a major change in deep water flows due to climate change. It is also necessary to store the waste with an adequate typology in order to eliminate any risk of criticality *in situ* [AMI 13].

Monitoring of a deep geological repository must be conducted over a long period of time and use relevant indicators. The IAEA has proposed indicators that are reported in Table 2.3. Depending on the compartments, the quantification of the various indicators will differ and will provide varied information on radionuclide inventory, concentrations, fluxes and transit times (Table 2.4).

Health effects	Direct safety indicators can be compared to existing standards (and natural background)
Risk	
Individual doses	
Concentration in the environment	Comparison with natural radioactivity
Inputs to the biosphere	Comparison with natural radionuclide fluxes
Flow through the different security barriers	
Waste	Radiotoxicity indicator based on the waste inventory

Table 2.3. *Excerpt from the IAEA hierarchy of safety indicators (source: [GRI 02])*

Indicator **Target compartment**	**Inventory in ????**	**Flow to (external inventory)**	**Concentration in**	**Transport time**
Type of waste	Waste	Waste		
Container	Precipitate	Waste packages	Waste packages	
Stamp	Stamp	Nearby		Stamp
Damaged area			Water in the damaged area	
Geosphere	Geosphere	Geosphere		Geosphere
Biosphere	Biosphere		Water of the biosphere	

Table. 2.4. *The various indicators that can be used in the different compartments (source: [STO 02])*

2.3.5.2. *Storage in boreholes*

Borehole disposal consists of placing the waste in vertical structures dug into the rock in order to isolate it from natural surface phenomena, to reduce the possibility of contact with humans by reducing their accessibility and finally to prevent the dispersion of their contents in the environment. Unlike deep geological repositories, all the operations of storage in boreholes are carried out from the surface, from the excavation and handling of the packages to the closure operations.

Three types of storage in boreholes can be mentioned according to the type of waste. The first type consists of immobilizing exothermic waste (such as vitrified waste or spent fuel) in a glassy gangue resulting from the fusion of the surrounding rock. The second type involves injecting liquid waste directly into the rock. The third type of disposal involves stacking solid waste packages in a borehole. France carried out burials in two wells (PS1 and PS3) in the Moruroa atoll following the atomic tests in the Pacific [MAR 07].

The immobilization of exothermic waste in molten rock has been explored in the United States [HEU 81]. There is or is not conservation of the container (capsule of diameter 50–100 cm).

Four concepts have been explored. The DUMP (Deep Underground Melt Process) concept, developed by the Lawrence Livermore National Laboratory in the 1970s, where the cavity is created by explosives or by nuclear tests between 2 and 4 km deep [SCH 78].

The DSB (Deep Self Burial) concept, initially developed by Sandia National Laboratories, consists of lowering to a depth of 2 km waste enclosed in high-density containers that are allowed to melt by the heat they give off. It was taken up again around 1990, notably in Russia, China and the United Kingdom, for the elimination of small quantities of HLW, but this time without fusion [CHE 13].

The DRD (Deep Rock Disposal) concept, also developed by Sandia National Laboratories [KLE 74], is a combination of the two previous concepts where liquids are injected directly into the rock.

The solidified waste in situ melting concept [ANG 76] consists of mixing solid waste with rubble inside a cavity. One of the most promising methods has been the burial of solidified, unconfined HLW in deep boreholes in the bedrock. The heat generated by the waste would melt the surrounding rock, and the molten rock/waste mixture would later resolidify into a low-solubility matrix as the heating rate decreased. The melting was stimulated by dolerite (a magmatic rock with very little glass). Although the rock cracked, there was no loss of fusion because the cracks were self-sealing [KLE 74]. Good fusion was obtained at a considerable depth of

2–5 km, in the shale, a silicate medium with very low permeability. The waste in dry, calcined or vitrified form is then lowered into the void space through the access hole, and the shaft is sealed. The energy released by radioactive decay raises the temperature to a point where the surrounding rock begins to melt. The waste is then dissolved there. The extent of this melting region grows until the heat generated is balanced by conduction away from the melting zone. Resolidification then begins and ends when radioactive decay has progressed to the point where the temperature falls below the melting point of the waste rock solution [SCH 78].

Similarly, self-destructing tungsten capsules could be used to dispose of heat-generating HLW tens of kilometers below the Earth's surface [OJO 12].

Effective sealing of a borehole after waste placement is usually required. However, the latter method has some disadvantages associated with natural materials, such as high melting temperatures, slow crystallization kinetics, the resulting sealing materials being generally porous with low mechanical strength, insufficient adhesion to the surface of the waste container and lack of flexibility for engineering controls. Yang *et al.* [YAN 19] show that natural granitic materials can be deliberately engineered through chemical modifications to improve the sealing ability of the materials for deep borehole disposal (DBD).

Only one option, derived from DSB, is still being studied in the United States [GIB 99, YAN 19]. The concept of *in situ* fusion as an ultimate waste disposal option shows that the placement of solidified HLRW in an *in situ* melting cavity with a crushed rock backfill not only eliminates the major shortcomings inherent in other melting alternatives, but also meets reasonable final disposal criteria [ANG 76]. Recent advances in the knowledge of continental crustal rocks and fluids at depths of several kilometers suggest that a much deeper repository could be a safer and more environmentally acceptable alternative to the HLRW problem [GIB 99]. For example, the nuclear test program carried out by France in the 1970–1990s in French Polynesia provided knowledge of the mechanisms of fracturing, rock fusion and trapping of radioactive substances in molten rock [IRS 19c].

Research and development activities at Sandia National Laboratories on the performance of a deep borehole for nuclear waste disposal have continued recently. These laboratories estimate the total cost of a borehole, and the placement of the 400 containers and final sealing of the borehole, at $40 million (2011) [BRA 12]. This type of radioactive waste disposal in deep boreholes can be implemented more quickly than in geological mine repositories because humanity has more experience in operating small deep boreholes than in large shallow mine-type boreholes. In addition, this disposal is likely to be less expensive and more flexible than mining disposal, while providing greater long-term isolation [BRA 17].

The concept of DBD for high-level nuclear waste has been around for about 40 years. Now, the US Department of Energy (DOE) is revisiting the concept through recent studies at Sandia National Laboratory and a field test. With DBD, nuclear waste will be placed in boreholes at depths of 3–5 km in crystalline basement rocks. However, the Nuclear Waste Technical Review Board (NWTRB) has raised concerns that the deep subsurface is more complicated, leading to science, engineering and safety issues [SCH 17].

Freeze *et al.* [FRE 19] prepared the safety case specific to deep drilling disposal of cesium and strontium capsules, but is generally applicable to other forms of waste, such as spent nuclear fuel. This package contains a safety strategy, an assessment basis and a safety assessment. The safety strategy includes management, implementation, design and assessment strategies. The assessment basis considers site selection, pre-closure and post-closure, which includes all barriers (waste, engineering, geosphere/natural barriers and biosphere) and the surface environment. The safety assessment includes a pre-closure safety analysis, a post-closure performance assessment and confidence-building analyses. For the latter, they use reactive transport codes to predict water movement after well closure. The development and placement of borehole seals over the waste placement zone are also important for confidence building in the design [FRE 19].

2.3.6. *Sending into space*

The principle of this option is to permanently rid the Earth of the most harmful radioactive waste by sending it into space beyond the atmosphere using spacecraft. Sending waste into space was mainly studied in the United States, by NASA, in the 1970s and early 1980s. This was envisaged for long-lived HLW from the reprocessing of spent fuel [SCH 74d].

Studies published by NASA in 1978 [BUR 78] consider five possible destinations for the waste (high earth orbit, lunar orbit, lunar landing, solar orbit, solar impact and solar system escape). The surface of the moon (reachable in a few days) and the orbit around the sun (in six months) were those considered the most interesting from a safety point of view. The option of storage on the moon was evoked, but its cost appeared prohibitive [BUR 78].

Sending into space would be made with the help of a launcher of heavy loads derived from the shuttle (HLLV, Heavy Lift Launch Vehicle) with a space shuttle and a high performance orbital transfer vehicle (OTV) [BUR 78].

NASA's projects were abandoned for reasons of cost and risks of failure of launch being too high. Other nations, such as the USSR and Kazakhstan, also considered this solution [IRS 19c].

The space disposal option was considered in the early 1980s as an adjunct because the total disposal of fuel rods from commercial power plants was not considered economically practical with the space shuttle technology. Space disposal of some HLW, however, may reduce calculated and perceived risks. The space disposal option combined with terrestrial disposal may provide a more flexible and lower risk overall waste management system [RIC 81].

2.3.7. *Immobilization in polar ice*

The immobilization of exothermic radioactive waste in the thick polar ice sheets of Antarctica or Greenland consists of placing the containers either on the ice or at a shallow depth, so as to cause them to sink progressively by melting the ice around them. Three options could be envisaged: either a surface repository and the package would progressively descend to the bedrock, or a mooring at mid-height with cables allowing for eventual retrieval and monitoring from the surface, or a temporary storage facility on the surface allowing for cooling of the waste, its retrievability and finally its burial by snow.

However, there are pockets of salt water in the snow which considerably accelerate the corrosion of the packages. Moreover, the ice is in perpetual movement which increases the risk of breakage of the packages and with climate change the permanence of the snow and ice is very random, if not zero [IRS 19c].

Philberth [PHI 77] estimates that the waste containers should be retrievable for a few centuries until future research has solved all the problems and ^{90}Sr and ^{137}Cs have lowered their activity to less than 0.1% (i.e. 300 years). Safe and very cheap retrievability can be guaranteed without docking the containers. The author calculates that the most active waste in the world for the next 30 years can be placed in 3.10^7 spherical containers, with a radius of 0.2 m and arranged over an area of 15 km radius and a depth of 20–100 m below the surface of Antarctica and Greenland. According to this author, the deposit would not affect the stability of the ice cap and that even the most unexpected natural instabilities in the ice caps and/or caused by climate change could not cause radioactive contamination [PHI 77].

Characterizing ice tectonic deformation, glacial erosion and sedimentation, and basal hydrologic conditions of ice caps is essential for selecting high-latitude nuclear waste disposal sites [IVE 12].

2.3.8. *Transmutation*

Transmutation consists of making a radioactive nucleus stable or decayed into another radioactive nucleus but with a shorter physical half-life. This is achieved by destroying the radionuclide by various operations, either by fission of a heavy nucleus using fast neutrons, or by capturing a neutron. To illustrate this, the example of the fission of a uranium-235 nucleus is presented. Following fission, the nucleus splits into two very radioactive fragments. One of the possible splits is shown in Figure 2.3, one with 143 nucleons and the other with 90 nucleons. These stabilize at the end of a cascade of disintegrations. The 143-nucleon fragment takes about 15 days to become stable, and the 90-nucleon fragment about 30 years because of strontium-90. The physical half-life of uranium-235 being 704 million years, transmutation allows an extremely important gain in time. It is thus possible to destroy plutonium isotopes and minor actinides by fission.

Separation has been explained above (section 2.2.5). It is an essential step as will be explained later.

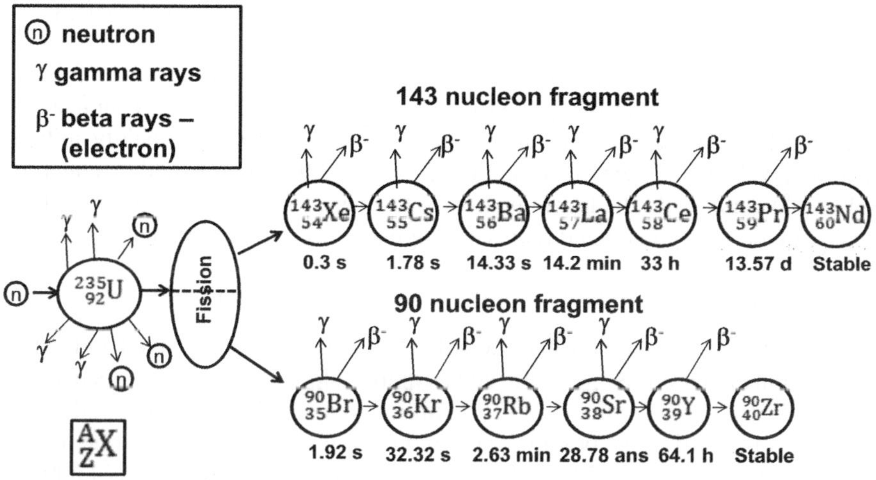

Figure 2.3. *Destruction by fission of a heavy nucleus ($^{235}_{92}U$) (X: symbol of the element; A: mass number; Z: atomic number) (source: [AMI 13])*

The theory is therefore attractive for destroying long-lived radionuclides. However, there are several important limitations. The two main ones are low yields and the risk of creating new radionuclides with even longer lives than the original radionuclides in the radioactive waste.

2.3.8.1. *A limited number of candidate radionuclides*

The radioactive elements selected for transmutation are those with a lifetime, measured by their half-life, of thousands or even millions of years. The advantage of an extremely long lifetime is that it is associated with low radioactive activity. On the other hand, this is offset by the disadvantage of the eternity it takes for them to disappear. The radioactive elements that are candidates for transmutation because of their long half-life are therefore minor actinides and certain fission products (^{90}Sr, ^{137}Cs).

Minor actinides, whose nucleus is heavier than that of uranium, are formed in reactors by the successive capture of one or more neutrons. They are generally alpha particle emitters, and these nuclei are radiotoxic if ingested. Minor actinides are fortunately not very mobile and not very abundant (less than 1 kg per ton of uranium at the exit of reactors). They are americium-243 and 241 (with physical half-lives of 7,367 and 432.6 years), curium-244 (18 years) and neptunium-237 (2,144,000 years). Plutonium, which is a major actinide, is not included in this list because it is already transmuted in some reactors. The destruction of actinides by fission reactions is attractive in theory but proves more difficult in practice [AMI 13].

Fission products that are very long-lived are few in number. They emit beta rays and are much less toxic if ingested than actinides. But, in certain chemical forms, they can be relatively mobile in storage conditions. These are mainly iodine-129 (period 16,100,000 years), cesium-135 (2,300,000 years) and technetium-99 (213,000 years). Only the transmutation of technetium-99 can be considered because it is sufficiently abundant and represents high activity and radiotoxicity [AMI 13].

2.3.8.2. *Very limited evidence of effectiveness and low yields*

According to Makhijani and Zerriffi [MAK 00], the long-lived radionuclides whose transmutation is possible and requires slow neutrons are technetium-99 (211, 000 years) and iodine-129 (16.1 million years). The long-lived radionuclides whose transmutation is possible and requires fast neutrons are americium-241 (432.6 years), neptunium-237 (2.1 million years) and plutonium-239 (24,000 years). For curium-244 (18 years) and tin-126 (100,000 years), transmutation is difficult. On the other hand, the long-lived radionuclides that have no possibility of transmutation are strontium-90 (29 years), cesium-137 (30 years), selenium-79 (60,000 years), cesium-135 (2.3 million years), zirconium-93 (1.5 million years), carbon-14 (5,700 years), chlorine-36 (300,000 years) and uranium-238 (4.5 million years).

Only the transmutation of technetium-99 has been demonstrated experimentally, becoming ^{100}Tc and, in a few minutes, stable ruthenium. The transmutation of minor actinides has been scientifically demonstrated. Subjected to neutron bombardment, minor actinides are effectively broken into lighter, stable or much shorter-lived nuclei. The transmutation of neptunium in a homogeneous medium with plutonium has been shown in the Petten reactor with high neutron flux. It has been confirmed that americium can be transmuted. Thus, an americium target was fissioned (thus transmuted) at 94% in the ECRIX experiment with the fast neutrons of the PHENIX reactor.

Transmutation yields are generally low. For example, in a powerful reactor that supplies 1 Gigawatt of electricity, neutrons transmute after three to four years of irradiation only 4.5% of the nuclear fuel into plutonium and fission products [CEA 12].

2.3.8.3. *Research for the distant future*

The IRSN and ASN issued an unfavorable opinion on transmutation on July 4, 2013 [ASN 13]. The ASN considers that the expected gains from transmutation of minor actinides in terms of safety, radiation protection and waste management do not appear to be decisive, particularly in view of the constraints induced on fuel cycle facilities, reactors and transport, which would have to use highly radioactive materials at all stages.

However, research has continued. The neutron, which penetrates nuclei without difficulty, is the ideal tool for transmutation, but it takes many neutrons to transmute waste. Nuclei are transmuted one by one and the smallest sample of matter contains billions of them. In practice, transmutation is carried out by bombarding the radionuclides with fast neutrons. Fast neutrons can be obtained by four options: in fast neutron reactors (FNRs), or with molten salt reactors operating with a mixed thorium–uranium-233 fuel, or with accelerators (ADS, accelerator-driven system), or with power lasers by the CPA method (Chirped Pulse Amplification) developed by Gérard Mourou, winner of the Nobel Prize in Physics in 2018. In 2020, none of these solutions was at the industrial stage. The FNR route is the most advanced with prototypes in several countries. However, many have experienced safety problems [AMI 21]. Several molten salt reactors with a fast spectrum have been tested, such as the American prototype MSBR (Molten Salt Breeder Reactor) or the French project of the CNRS (Molten Salt Fast Reactor) [CEA 15]. For ADS, research is underway mainly in China (CI-ADS, Chinese Initiative-ADS) [YAN 17], in Europe (Myrrha at SCK.CEN in Mol), in India, in Japan and in Sweden. For power lasers, only two patents have been filed [TAJ 17, TOS 18].

In molten salt reactors, the actinides introduced into the salts remain there until they are completely burned. The ADS have a higher performance in principle. While FNRs are able to burn their own actinides by recycling them, ADSs are able to incinerate a core made up solely of actinides. However, the first demonstrators or prototypes, such as the European project MYRRHA (Multi-purpose Hybrid

Research Reactor for High-tech Applications), are envisaged at best for the middle of the next decade [ANO 20a].

There are still many technological obstacles to overcome before these systems can be implemented. The results obtained with the available tools, the Petten high neutron flux reactor in Holland and the PHENIX reactor with fast neutrons, have confirmed that the transmutation of fission products is not very profitable.

2.3.8.4. *Limited applications*

To perform an efficient transmutation, the separation of radionuclides is mandatory to create a target; otherwise, there is a significant risk that long-lived radionuclides are created by neutron capture. Worldwide, there are proven solutions for each of the routes studied. Thus, the separation of americium and curium can be done by the DIAMEX-SANEX process, the grouped separation of actinides (Pu+AM) by the GANEX process and the separation of americium alone by the EXAm process [CEA 12].

The transmutation of minor actinides can only make sense in conjunction with a plutonium recycling strategy. In order to transmute minor actinides, a technology capable of using the major actinide, plutonium, in a perennial way is required. FNRs are capable of meeting these two objectives. Transmutation can only be envisaged in the framework of a future nuclear fleet comprising reactors adapted to transmutation and cycle installations capable of ensuring the separation and recycling of radionuclides of interest. This strategy does not therefore apply to the glass packages of the current nuclear power plant (heterogeneous radioactive waste already produced and committed waste). This final waste is intended for geological disposal.

In any case, the transmutation of minor actinides is a complex and slow process, and obtaining high performance requires several passes in the reactor and, consequently, a multi-recycling of the elements considered, leading to multiple and costly manipulations of dangerous materials [CEA 12].

2.3.8.5. *Significant additional costs*

Transmutation will only make sense if it is implemented over long periods (typically several centuries) and if it is accompanied by an "end-of-life" strategy, the duration of which can be estimated today at between one and two centuries, in order to sufficiently reduce the final inventories of actinides sent to geological storage.

Transmutation has impacts on inventories, on waste (reduction of radiotoxicity by a factor of 20–100), on the cycle's facilities (obvious problems of radioprotection and thermics in fuel fabrication, transport and processing operations, and this is due especially to actinides), and on the safety of supply (transmutation makes the whole cycle more complex). Transmutation also has an impact on the economy with an additional cost of 4–9% if FNR reactors are used and an additional cost of 25% if ADS is used. These different observations suggest that the disadvantages of transmutation of curium would greatly exceed its advantages [CEA 12].

2.4. Conclusions

The disposal methods of radioactive waste are numerous and depend strongly on the activity level of the waste and the half-life of the radionuclide. Figure 2.4 summarizes these various theoretical solutions.

For very short-lived radioactive waste, the solution of waiting for the effect of physical decay is ideal and relatively easy to implement and manage. For VLLW and LLW wastes, surface and subsurface storage is a good solution. The waste can be immobilized in various matrices (cement, bitumen, glass, etc.).

On the other hand, for intermediate- and high-level long-lived wastes, perfect solutions do not exist and, generally speaking, the solutions considered are deep geological burial or drilling at great depth, associated with multiple artificial and natural barriers.

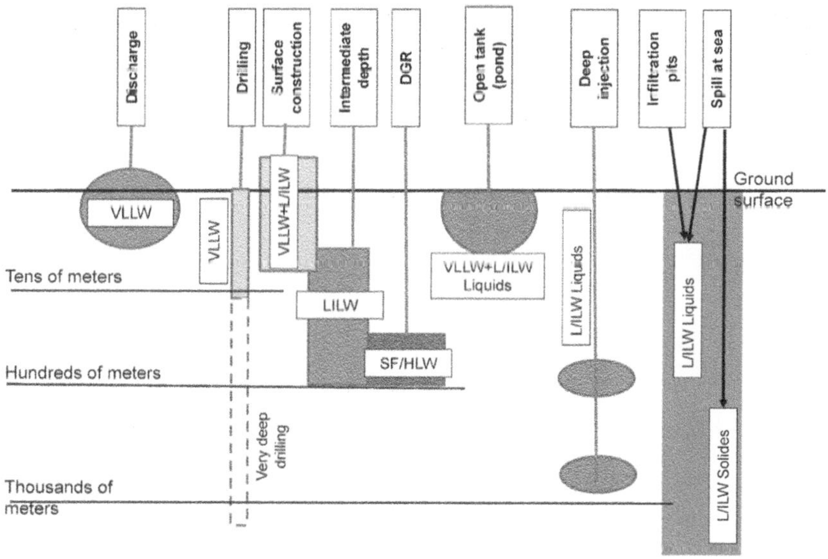

Figure 2.4. *The various solutions for managing radioactive waste and spent nuclear fuel (source: [OJO 14]). DGR: deep geological repository; HLW; high-level waste; L/ILW: low/intermediate-level waste; SF: spent fuel; VLLW: very low-level waste. For a color version of this figure, see www.iste.co.uk/amiard/radioactive.zip*

The safe disposal of radioactive waste, especially spent fuel, former military fissile material and other forms of HLW, is one of the major challenges facing contemporary science. At present, the preferred solution at the international level is geological disposal by burial in a multi-barrier mining and artificial repository [GIB 99].

3

Management of Historic Radioactive Waste and Low-level Waste Around the World

3.1. Introduction

The management of very low-level and low-level radioactive waste is under control in most nuclear countries. For the nuclear power industry, it is important to demonstrate that all nuclear waste, even the most highly radioactive, can, in theory, be isolated from humans and the environment in complete safety for as long as necessary. Therefore, the development of techniques for the treatment, conditioning and final disposal of LLW and ILW is ongoing. In many countries, public pressure has given a strong impetus to progress in radioactive waste management and this can be expected to continue in the future.

The volume of radioactive waste will only increase in the coming years, both in developing and industrial countries, due to the continuous development of nuclear energy and especially the decommissioning and dismantling of decommissioned reactors.

With respect to the day-to-day handling and management of radioactive waste, developing and industrial countries face different problems and therefore have different immediate needs. International cooperation and exchange between national programs and international organizations therefore play a very useful role for all interested parties [ZHU 89].

For LLW most countries have chosen the solution of treatment, then compaction and storage, with partial incineration also existing. The storage system is generally a surface or shallow installation, or subsurface, with a cover.

3.2. Management of historical radioactive waste

Radioactive waste is as old as the first research on radioactivity. At the beginning, little attention was paid to its storage because its dangerousness was not known. Thus, historical waste is characterized by random storage, not very safe. The principal historical wastes concern those from uranium mining activities, from health applications, from the military and from the watchmaking industry. In the early days, direct discharge of waste into the environment (rivers, lakes, soil) and dumping at sea were the rule. These wastes are generally of low activity but some are long-lived. However, few states have storage centers for this type of waste.

3.2.1. *Uranium extraction and concentration waste*

The nuclear fuel cycle begins with mining operations to extract uranium (or thorium) ore and preconcentrate the uranium. These various operations are subject to contamination of the environment by direct atmospheric discharges (dust, radon, etc.) and liquid discharges with mine water, not to mention the emissions from the ores or waste rock.

All countries that have exploited uranium mines have waste rock deposits that are still relatively rich in radioactivity. This is particularly true for Canada and Australia. The NEA [AEN 02b] has identified the national policies of the various countries that have exploited uranium mines and that produce uranium (Argentina, Australia, Brazil, Canada, Czech Republic, Egypt, Finland, France, Gabon, Germany, Hungary, Japan, Kazakhstan, Portugal, Romania, Russia, Spain, Sweden, Ukraine, United States and Uzbekistan). In particular, this document provides the various storage sites for tailings from mining operations.

The main types of waste generated by the extraction and processing of uranium ore are residues and waste rock. Residues are the waste products produced by the milling of ore and the chemical concentration of uranium, often in the form of sludge. When dry, residues have the consistency of fine sand. Waste rock is simply rock removed during excavation to access the ore. Waste rock contains little or no uranium and has little economic value [CCS 20].

During uranium mining, the radiation doses from mining waste are due to three radionuclides: natural uranium, radium 226 and thorium 230. The doses are, respectively, less than 2, 5 and 10 $\mu Gy.h^{-1}$ for the waste rock from mining operations, the plant residues and contaminated equipment, and the decontamination waste [AER 07b]. The radiation concentrations and doses for the various stages of thorium mining processes are compiled in Table 3.1.

Solid waste	Alpha emitters ($Bq.g^{-1}$)	Beta emitters ($Bq.g^{-1}$)	Radiation dose ($mGy.h^{-1}$)
Extraction residues	0.5	5	0.5
Residue from preconcentrations	0.8	4	0.4
Residues from the separation of ores	80	300	50

Table 3.1. *Typical activities of solid mine waste and thorium mining treatments (source: [AER 07b])*

Typical solid wastes from uranium mine wastes are presented in Table 3.2. At uranium mines, the typical liquid effluents from the effluent treatment plant are listed in Table 3.3. The effluents from conventional mines represent 1–8 $kBq.m^{-3}$ for ^{222}Rn [AER 07b].

Solid waste	Alpha emitters ($Bq.g^{-1}$)	Beta emitters ($Bq.g^{-1}$)
Mining waste	15	10
Retention pond waste	50	10

Table 3.2. *Typical activities of solid mine wastes after chemical treatments for uranium extraction (source: [AER 07b])*

Liquid waste	Alpha emitters ($Bq.L^{-1}$)	Beta emitters ($Bq.L^{-1}$)	Natural U ($Bq.L^{-1}$)	Radium-226 ($Bq.L^{-1}$)
Mine water	1–10	1–10	1–5	0.5–1.0
Crushing washing, grinding	-	-	2.5–10	0.2–0.5
ETP discharge	0.5–2.5	0.5–1.5	0.02–0.2	0.01–0.03

Table 3.3. *Typical activities of liquid mine wastes during uranium extraction separation treatments (source: [AER 07b]). ETP: effluent treatment plant*

Typical liquid wastes from thorium mining and from the various phases of ore processing contain high levels of alpha and beta emitters, including alpha-emitting radium-226. The same is true when treating monazites (Table 3.4).

Liquid waste	pH	Alpha emitters (Bq.L^{-1})	Beta emitters (Bq.L^{-1})	^{226}Ra (Bq.L^{-1})
Monazite treatments				
Acidic effluent	1.6–2.0	100–300	400–600	150–200
Alkaline effluent	12–13	600–900	900–1,000	300–400
Thorium treatment				
Non-nitrated effluent	1–3	20–30	100–110	25–40
Nitrated effluent	1–3	20–70	200–700	25–30

Table 3.4. *Characteristics of liquid mining waste from monazite and thorium ore processing (source: [AER 07b])*

3.2.2. *Direct discharges of liquid wastes into waterways and reservoirs*

Examples of direct discharges into a river are primarily US discharges into the Columbia River and Soviet discharges into the Techa. The Hanford site in the United States borders the Columbia River. This site had nine plutonium reactors and five reprocessing plants. A significant part of the liquid LLW and ILW was discharged into this river as early as 1944. From March 1949 to August 1951, the plutonium reactors at Mayak had only one cooling loop in direct connection with the waters of Lake Kyzyltash (fed by the Techa), which consequently became heavily contaminated. Similarly, direct discharges of low and medium activities were made into nearby reservoirs (Metlinski Pond, etc.). Then, the discharges were made into a closed water reserve, Lake Karachay. These two sites, American and Soviet, were therefore formerly very polluted by numerous radionuclides [FON 60]. This was verified by Aakrog *et al.* [AAK 97] in the entire southern Urals where they took 25 samples of contaminated soil as well as samples of grass and litter.

Radioactive discharges into the Techa River were 100 PBq, including 10 PBq for ^{90}Sr and ^{137}Cs, respectively, between 1949 and 1956 [AAK 03]. The river contained in its sediments in 1990 0.3 TBq of ^{90}Sr, more than 6 TBq of ^{137}Cs and about 8 GBq of ^{239,240}Pu. At that time, sediments, plants and fish from the Techa River were heavily contaminated with ^{60}Co, ^{90}Sr, ^{137}Cs and plutonium for a portion of the river from 50 km downstream of the discharges to its confluence with the Iset River 280 km away. Contamination was limited to the top 10 cm of sediment. The dose rates reached 16, 15 and 28 μGy.h^{-1} at the surface, on the bottom and on the banks of the river, respectively [TRA 93].

However, the principal direct releases into the environment were a result of the atmospheric testing of atomic bombs [AMI 18].

During normal operation, nuclear reactors released into the atmosphere mainly noble gases (argon, krypton and xenon), as well as tritium. Thus, PWR and BWR reactors released annually 3 TBq.GWe^{-1}, HWR 500 TBq.GWe^{-1}, GCR and FBR, respectively, 9 and 100 TBq.GWe^{-1}. Annual releases of ^{14}C are 120, 450, 4,800 and 540 GBq.GWe^{-1} for the PWR, BWR, HWR and GCR reactors, respectively. Releases of ^{131}I are, respectively, 0.93, 1.8, 0.19, 1.4 and 14 GBq.GWe^{-1} for PWRs, BWRs, HWRs, GCRs and LWGRs.

The annual liquid releases of tritium from the PWR, BWR, HWR, GCR, LWGR and FBR reactors are 23, 0.62, 377, 120, 11 and 3 TBq.GWe^{-1}, respectively. Many fission and activation products (^{51}Cr, ^{54}Mn, ^{55}Fe, ^{59}Fe, ^{58}Co, ^{60}Co, ^{65}Zn, ^{110m}Ag, ^{134}Cs and ^{137}Cs) are also directly discharged into the aquatic environment.

The reprocessing of irradiated fuel, and in particular the Sellafield plant in the United Kingdom, has released large quantities of radionuclides. Direct releases to the atmosphere are 12,300,000, 41,000, 2, 5.7 and 1.5 GBq.GWe^{-1} per year for ^{85}Kr, ^{3}H, ^{14}C, ^{129}I and ^{137}Cs, respectively. Liquid releases are 643, 39, 36, 13, 0.54 and 0.032 TBq.GWe^{-1} per year for ^{3}H, ^{106}Ru, ^{90}Sr, ^{137}Cs, ^{14}C and ^{129}I, respectively.

UNSCEAR has calculated that radioactive tracers used in medicine and industry have introduced into the environment 90, 30, 19, 30, 180 and 1600 GBq of ^{3}H, ^{14}C, ^{85}Kr, ^{125}I, ^{131}I and ^{133}Xe per million inhabitants, respectively [AAK 96].

Nuclear accidents have also introduced large quantities of radionuclides. Thus, the graphite fire accident in the Windscale reactor in 1957 released into the atmosphere approximately 0.0–1 TBq of ^{131}I, 22–94 TBq of ^{137}Cs, 5 TBq of ^{89}Sr, 0.22 TBq of ^{90}Sr, 8.8 TBq of ^{210}Po, 5 TBq of ^{3}H and 0.0016 TBq of ^{239}Pu. Accidents also occurred in radioactive effluent tanks, such as the one at Kyshtym, where 74 PBq were released into the environment (mainly short-lived radionuclides such as ^{95}Zr and ^{144}Ce, as well as 2 PBq of ^{90}Sr and 30 PBq of ^{137}Cs). In addition, radioactive deposits have been created around the Mayak plant. One of these, Lake Karachay (0.45 km^{2}), suffered a violent windstorm in the summer of 1967 and about 20 TBq of ^{90}Sr and ^{137}Cs were dispersed over a vast geographical area of 1,800 km^{2}. The losses of American and especially Soviet submarines have been numerous. One of the last, concerning the submarine Komsomolets, occurred in April 1989. Its reactor contained 2.8 PBq of ^{90}Sr, 3.1 PBq of ^{137}Cs and 16 TBq of plutonium. The Americans lost two atomic bombs in Palomares in Spain (0.1 TBq of ^{239,240}Pu) and in Thule in Greenland (1 TBq of ^{239,240}Pu, 0.02 TBq of ^{238}Pu and 0.1 TBq of ^{241}Am) [AMI 18].

The losses of sealed sources have been numerous. The two major accidents concerned Mexico City (Mexico) in 1983 (16.7 TBq ^{60}Co) and Goiânia (Brazil) in 1987 (50.9 TBq ^{137}Cs) [AMI 19]. The two principal losses of satellites containing radioactive sources were that of the American satellite SNAP-9A in 1967 (0.33 TBq ^{238}Pu) and the Soviet satellite Cosmos 954 in 1978 (3 TBq ^{90}Sr, 0.2 TBq ^{131}I and 3 TBq ^{137}Cs) [AAK 96].

3.2.3. *Historical military waste*

The first use of nuclear energy was military with the manufacture of atomic bombs. The American Manhattan project was initiated in wartime, and the protection of humans and the environment were not priorities. Therefore, all the sites of this project were the location of radioactive waste deposits with little precaution. This resulted in a high level of radioactive contamination of the sites, notably Hanford and Oak Ridge, which was detailed in the fourth volume of the "radioactive risk" series [AMI 21], because it led to extensive clean-up and dismantling programs. All these American sites still have enormous quantities of liquid and solid radioactive wastes.

It was the same on the Soviet side at all its military sites that were kept secret for a long time. Nuclear accidents have also occurred at several of these sites, contaminating the environment. To a lesser extent, the same is true for the other nuclear powers.

In the same way, the sites of atomic bomb tests still represent contaminated zones, frequently containing deposits of radioactive waste that are summarily confined. This is, for example, the case for the Marshall Islands, New Zealand, Mururoa and Fangataufa [AMI 18].

A special case is the very large fleet of Soviet submarines. Out of a total of 254 submarines, a large number (120) have been disarmed and stored without much surveillance in various ports. In particular, 96 submarines are stored with their spent fuel still in place [SAW 01]. Their dismantling was carried out with the help of various international cooperation. A declaration was made in September 1996 between Norway, Russia and the USA of the AMEC (Arctic Military Environmental Cooperation) to manage the spent fuel and radioactive waste of the Russian Northern Fleet [SAW 00]. Russian radioactive waste in the Northwest is considerable, with 45,500 m^3 of liquids, 26,200 m^3 of solid waste and 82,400 spent fuel assemblies [SAW 01]. To this must be added the numerous Soviet submarines that sank, sometimes with their crews, as well as their nuclear reactors and atomic weapons [AMI 18]. Surveillance is carried out by the Norwegian Radiation and Nuclear Safety Authority (DSA).

3.2.4. *The ancient uses of radium*

Very soon after the discovery of radium, exciting applications were made in many fields. The applications were particularly varied in the fields of hygiene and beauty, as well as in that of pharmacy. Radium was seen as a symbol, a myth or a "magic potion".

3.2.4.1. *The use of radium in hygiene and pharmacy*

Thus, remedies, creams, ointments, radium fountains producing radon-laden water, razor blades, fishing rods, baits, dishcloths and wool were manufactured, enriched with radium and/or thorium.

In August 1937, the Codex Commission listed thorium and radium in Table A of poisonous substances, which specified that these radioactive substances *could only be prescribed by a physician's prescription and could only be sold or dispensed to the public upon presentation of that prescription* (Zerbib, 2020; personal communication).

3.2.4.2. *Waste from the watchmaking industry*

For a long time, the hands and even the numbers on dials, watches and clocks were made luminous by the application of radium or tritium. The first factories were located in the United States. In 1915, in Newark, New Jersey (USA), the Radium Luminous Material Corporation, which later became the US Radium Corporation, was the first to use luminescent paint (based on zinc sulfide to which a small amount of radium was added). This company employed hundreds of female workers (Radium girls). Another factory, Radium Dial, was located in Ottawa, Illinois. In Europe, the watch industry using radium was located mainly in Switzerland (Biel and La Chaux-de-Fonds) and in eastern France (Besançon).

The luminescent paint was applied with a fine brush whose tip was thinned between the women's lips, causing oral contamination that triggered cancers. Some of the highly radioactive residues have been left near these long-gone workshops [MUR 16].

3.2.5. *Submergence in the ocean floor*

Historically, in order to isolate radioactive waste from the human environment, it was immersed as early as 1946, and for four decades, in the ocean floor by most nations using nuclear energy. The most dangerous were conditioned. The great dilution of the marine environment and the long period of isolation of the packages were supposed to be sufficiently safe to protect humans from the dangerousness of this waste.

The first regulatory texts did not appear until 1958 with the United Nations conference on the law of the sea and in 1961 with the recommendations of the IAEA on the choice of sites and the supervision of a responsible body. Thus, as early as 1967, the European countries dumped radioactive packages under the responsibility of the NEA, which is dependent on the OECD. In 1972, the London Convention prohibited the dumping of high-level radioactive waste and required a special authorization for low-level radioactive waste. This convention took effect in 1975. Several countries adopted a voluntary moratorium on such dumping in 1983, which 10 years later resulted in a total ban on such dumping.

From 1946 to 1993, different types of radioactive waste were immersed. These were liquid wastes, directly discharged into the sea or put in liquid form in containers, and solid wastes, for the most part conditioned in metal drums after incorporation in concrete. The United States and the USSR have also sunk nuclear reactors, some of which still contained spent fuel.

Fourteen countries have conducted dives at more than 80 sites in the Pacific and Atlantic with a total activity of 85,000 TBq (Table 3.5). The vast majority of the radioactivity (99%) is represented by beta and gamma emitters (^{90}Sr, ^{137}Cs, ^{55}Fe, ^{58}Co, ^{60}Co, ^{125}I, ^{14}C and tritium).

Following the work of the Grenelle de la Mer, a French governmental program, it was decided to set up a better monitoring and a more efficient control of the marine environment. For submerged radioactive waste, the priorities were to consolidate the inventory of underwater dumps of this type of waste. Two reports have been published [IAE 99, MIN 15].

Geographical area	Activity (TBq)	Activity β and γ (TBq)	Total activity (TBq)	Percentage of total activity (%)
Northeast Atlantic	675	41,645	42,320	49.7
Northwest Atlantic	-	2,942	2,942	3.5
Arctic	-	38,370	38,370	45.1
Northeast Pacific	0.04	554	554	0.7
Western Pacific	-	892	892	1
Pacific (French Polynesia)	0.07	0.02	0.09	-
Total	**675**	**84,000**	**85,100**	**100**

Table 3.5. *The distribution of radioactivity of submerged wastes in various geographical areas (source: [IAE 99])*

3.2.5.1. *European immersions*

For the European States, the principal divestitures (1,800 TBq) in the Northeast Atlantic were carried out from 1949 to 1966 by the United Kingdom, and to a lesser extent by Belgium (Table 3.6).

Site	**Latitude**	**Longitude**	**Depth (m)**	**Date**	**Tonnage (t)**	**Activity (TBq)**
1	48 30’N	13 00’W	3,600–4,000	1949	9	0.04
2	49 50’N	2 18’W	65–160	1950–1963 (1/year)	17,274	60
3	55 20’N	11 20’W	2,700	1951	33	0.2
4	55 80’N	12 10’W	2,800	1953	57	0.15
5	32 37’N	14 50’W	4,000–4,200	1955	1,453	1.7
6	32 42’N	19 30’W	3,600–4,100	1957, 1958	7,098	131
7	32 38’N	20 50’W	2,100–4,800	1961	4,360	81
8	46 27’N	6 10’W	4,200–4,600	1962	253	6.7
9	45 27’N	6 16’W	4,100–4,800	1963, 1964	10,201	850
10	48 20’N	13 16’W	1,900–4,500	1965, 1966	2,803	617
Total					**43,500**	**1,800**

Table 3.6. *Immersions in the Northeast Atlantic from 1949 to 1966 (source: [IAE 99]). All the immersions were carried out by the United Kingdom, except at sites 2 and 9 where Belgium joined the United Kingdom*

After 1996, the 1967 (293 TBq), 1969 (834 TBq) and 1971–1982 (35,882 TBq), disposal campaigns were coordinated by the NEA (Tables 3.7–3.9). On these occasions, the NEA established a monitoring program (1981–1995) for its disposal site (CRESP, Co-ordinated Research and Environmental Surveillance Programme). Until 1988, the examination of samples of sea water, sediments and marine organisms taken from the different sites did not reveal any concentration of radionuclides greater than that due to the fallout from nuclear weapon tests, except in some cases where higher concentrations of cesium and plutonium were measured in samples collected in the immediate vicinity of the packages [CAL 89, NEA 96]. European dumping took place at 13 different sites.

Country	Number of containers	Mass (t)	Activity (TBq)
Germany	480	181	0.2
Belgium	1,945	600	7
France	31,596	9,184	220
United Kingdom	-	722	66
Netherlands	-	207	0.07
Total	**34,021**	**10,894**	**293.27**

Table 3.7. *Coordinated immersions at the 1967 NEA site (400 km from Galicia, Spain, at 4,600 m depth)*

Country	Number of containers	Mass (t)	Activity (TBq)
Belgium	2,222	600	18
France	14,800	5,015	134
Italy	100	45	0.2
Netherlands	-	303	1
United Kingdom	-	1,878	665
Sweden	2,895	1,081	3.2
Switzerland	100	224	13
Total	**20,117**	**9,146**	**834.4**

Table 3.8. *Coordinated immersions at the 1969 NEA site (900 km from Brittany, France, at a depth between 4,000 and 4,600 m)*

Country	Number of containers	Mass (t)	Activity (TBq)
Belgium	51,157	27,026	2,090
Netherlands	-	23,788	29,050
United Kingdom	>28,428	18,652	335
Switzerland	7,370	5,097	4,407
Total	**86,955**	**74,563**	**35,882**

Table 3.9. *Coordinated immersions at the NEA site between 1971 and 1982 (Bay of Biscay, 1,000 km from the French coast)*

In addition, the United Kingdom carried out several immersion campaigns between 1949 and 1982 in its territorial waters, in particular the Irish Sea (10 TBq), in 1968 and 1970 (13,500 TBq).

3.2.5.2. *American immersions*

The United States immersed in the Atlantic Ocean from 1949 to 1967, 34,282 containers with a total activity of $2.94.10^6$ GBq distributed over 11 different sites. The country's immersions in the Pacific Ocean were distributed over 18 sites during the period 1946–1970. The number of containers immersed was 56,261 and the total activity was $5.54.10^5$ GBq [IAE 99].

3.2.5.3. *The Soviet immersions*

The immersions of the Soviet Union are very numerous. Liquid radioactive waste was dumped at five sites in the Arctic Ocean from 1959 to 1991 for a total of 190 334 m^3 and an activity of 764 TBq.

Solid waste of low and medium activities was immersed in the Arctic seas at 11 sites near New Zealand in 6,508 containers with a total activity greater than 585 TBq. In addition, the USSR immersed several objects, including six reactors with spent fuel still in place at four sites near New Zealand between 1965 and 1981. This represents an activity of 36 876 TBq. To these deposits should be added 10 reactors without their spent fuel at the same sites, immersed from 1965 to 1988 (activity 143 TBq). The Soviet Union dumped 123,497 m^3 of liquid waste (456 TBq) at nine sites in the Pacific Ocean from 1966 to 1992. From 1968 to 1992 at four sites in the Pacific Ocean, the USSR dumped 6,642 containers and 39 ships with a total activity of 418 TBq [IAE 99].

In 1992, the Russian Federation dumped in the Barents Sea and in the seas of the eastern coast of the Pacific Ocean (notably the Sea of Japan) various liquid and solid nuclear wastes (activity 2 TBq) and in 1993, still in the Sea of Japan, liquid wastes contaminated mainly by ^{137}Cs, ^{90}Sr and ^{60}Co (13.9 GBq) [IAE 99].

A new assessment of the Novaya Zemlya disposal sites was conducted in 2003/04. Levels of ^{137}Cs contamination in Stepovogo and Abrosimova bays were similar to background environmental levels and there was no detectable leakage of radionuclides into Tsivolki Bay. Levels of ^{137}Cs contamination in near-surface sediments in 2003/04 were 1–11 $Bq.kg^{-1}$, 4–268 $Bq.kg^{-1}$ and 13–20 $Bq.kg^{-1}$ in Tsivolki, Stepovogo and Abrosimova bays, respectively [DAH 09].

3.2.5.4. *Immersions by other nations*

Japan dumped 3,031 containers at six different sites (close to the Japanese coast) between 1955 and 1969 for a total activity of $1.51.10^4$ GBq. Switzerland dumped from 1969 to 1982, 7,420 containers with an activity of $4.42.10^6$ GBq at three sites in the Atlantic. New Zealand dumped in four different sites in the Pacific Ocean, close to the New Zealand coast, from 1954 to 1976 a total activity of $1.04.10^3$ GBq. North Korea dumped 115 containers from 1968 to 1972 at a single site near the Korean peninsula. The present activity is unknown. Sweden dumped in a single operation in 1969, 2,895 containers representing an activity of $3.24.10^3$ GBq in the Atlantic Ocean. In addition, this same country dumped 14.8 GBq in the Baltic Sea in 1959 and 1961 [IAE 99]. Apparently Canada has not carried out any dumping operations if one believes the IAEA.

The global balance of ocean dumping is provided in Table 3.10.

Type of waste	Atlantic	Pacific	Arctic	Total	Percentage of total activity
Reactors with fuel	0	0	36,876	36,876	43.34
Fuel-less reactors	1221	166	143	1,530	1.80
Solid waste from low activity	44,042.5	828.9	585.4	45,448.8	53.42
Liquid waste from low activity	<0.01	458.5	764.7	1,223.2	1.44
Total	45,263.5	1,445.4	38,369.1	85,078.0	
Percentage of total activity	53.20	10.70	45.10		100.00

Table 3.10. *Activity distribution (TBq) for different types of wastes dumped in the various oceans (source: [IAE 99])*

In the Atlantic Ocean, the United Kingdom accounts for 77.5% of the deposits, Switzerland for 9.8%, the United States for 6.5% and Belgium for 4.7%. The immersions of Germany, the Netherlands and France are very low. In the Pacific, the USSR's deposits represent 59.5%, those of the United States 38.9%, Japan 1.2% and New Zealand 0.3%. In the Arctic Ocean, all the deposits are due to the USSR [IAE 99].

3.3. International recommendations of the IAEA and NEA

Two organizations that have worked and published extensively at the international level to help various nations manage their radioactive waste are the IAEA and the NEA, an agency of the OECD.

3.3.1. *General recommendations*

The IAEA published in 2009 [IAE 09b] 22 requirements for the proper management of nuclear waste. These requirements for the management of radioactive waste fall into 10 main categories.

From a legal, regulatory and policy framework perspective, the government must provide an appropriate national legal and regulatory framework within which radioactive waste management activities can be planned and conducted safely. This includes allocating responsibilities, securing financial resources and establishing independent regulatory functions. The protection of neighboring states must also be taken into account. To this end, the government must ensure that a national policy and strategy for radioactive waste management is established. This policy and strategy must be consistent with fundamental safety principles and not contradict international conventions and codes ratified by the State. The State must designate a regulatory body that is responsible for this policy. This body shall review and evaluate the safety case and environmental impact assessment for radioactive waste management facilities and activities, as prepared by the operator both prior to licensing and periodically during operation. It provides for the issuance, modification, suspension or revocation of licenses. Enforcement action must be taken in the event of deviation or non-compliance with requirements and conditions by the operator.

The operator is responsible for the safety of radioactive waste management facilities or activities. He must develop a safety case and must ensure that the activities necessary for site selection, design, construction, commissioning, operation, shutdown and decommissioning are carried out in accordance with legal and regulatory requirements.

Measures must be implemented to ensure an integrated approach to safety and security in the management of radioactive waste prior to disposal. The interdependencies between the various stages of management shall be duly taken into account. Management systems shall be applied for all stages and elements of the pre-disposal management of radioactive waste.

The production of radioactive waste must be known and controlled. Radioactive waste emissions must be reduced to the minimum possible. Radioactive waste must be characterized and classified.

The treatment of radioactive waste, i.e. all radioactive materials without use, must be carried out. The treatment is carried out according to the different stages of their management (pre-treatment, treatment, packaging, transport, storage and disposal). The waste packages must be designed to contain the waste even in the event of an accident during handling, storage, transport and disposal of the waste.

The storage of radioactive waste must be designed in such a way that it can be inspected, monitored, retrieved and stored under conditions suitable for its subsequent management.

Waste packages and unpackaged waste accepted for treatment, storage and/or disposal must meet criteria consistent with the safety case.

The operator must prepare a safety case and a safety assessment, possibly updated if the management program changes. This safety case shall be complete and shall demonstrate the level of protection. The documentation must be clearly written and include arguments justifying the approaches taken in the safety case based on traceable information. An independent review and approval of the safety case is performed by the regulatory body. The operator conducts periodic safety reviews and implements any safety improvements required by the regulatory body.

The development of radioactive waste management facilities is the responsibility of the operator. The choice of site and design of the facilities must ensure safety during their lifetime. The construction and commissioning of the facilities must follow the design described in the safety case and approved by the regulatory body. Specific checks shall be carried out during commissioning. The operation of the facility shall follow national regulations and conditions imposed by the regulatory body and shall be based on documented procedures. Special attention must be paid to the maintenance of the plant to ensure its safe operation. Emergency preparedness and response plans are submitted to the regulatory body for approval. The operator shall develop, at the design stage, an initial plan for the shutdown and decommissioning of the radioactive waste management facility prior to disposal and update it periodically throughout the operating period. In addition, assurance shall be provided that sufficient funds will be available to carry out shutdown and decommissioning.

Finally, there are two last requirements concerning the system of accounting and control of nuclear materials for certain facilities subject to this rule. For existing facilities, security must be reviewed for compliance with the requirements, with possible upgrades performed by the operator [IAE 09b].

Specific safety requirements for the disposal of radioactive waste have been issued by the IAEA [IAE 11a]. They are 26 in number and partly repeat the 2009 requirements and complete them.

Key requirements are listed as the responsibilities of the government, the regulator and the operator. The document emphasizes passive means for disposal facility security and understanding of a disposal facility and confidence in security. Clarification is provided that security is achieved through multiple security functions (containment and isolation of waste, performance of these physical barriers).

Monitoring and control of passive safety devices must be applied to protect and preserve passive safety devices. The IAEA recommends a step-by-step development and evaluation of disposal facilities.

Several requirements address the preparation, approval and use of the safety document and safety assessment for a disposal facility. The safety file and supporting safety assessment must demonstrate the level of protection for people and the environment and provide assurance to the regulatory agency and other interested parties that the safety requirements will be met. This safety file shall be documented to a level of detail and quality sufficient to inform and support the decision to be made at each stage and to permit an independent review of the safety case and supporting safety assessment.

Requirements 15 through 19 address site characterization for a disposal facility, its design, construction, operation and closure. They largely repeat the recommendations of the 2009 document.

Waste packages and unpackaged waste accepted for placement in a disposal facility must meet criteria that are fully consistent with and derived from the safety file for the operating and post-closure disposal facility.

A monitoring program should be conducted before and during construction and operation of a disposal facility and after shutdown. After shutdown, institutional controls must be maintained.

The latest requirements for the national system of nuclear material accountancy and control, requirements for nuclear security measures, management systems and existing disposal facilities are similar to those in the 2009 text [IAE 11a].

For the storage of spent nuclear fuel, the IAEA has published a specific safety guide [IAE 12a]. The same is true for the preparation of the safety case and the safety assessment for the disposal of radioactive waste; the specific safety guide was published in 2012 [IAE 12b].

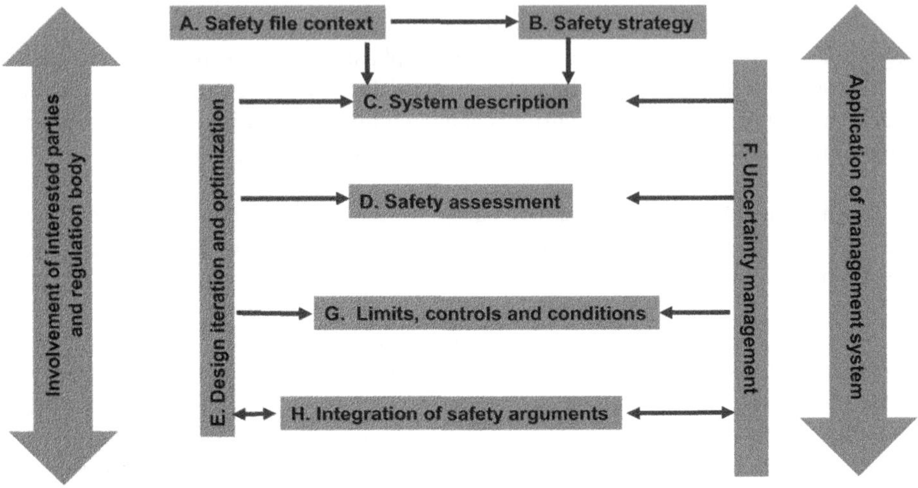

Figure 3.1. *Application of the management system and the process of interaction with the regulator and interested parties (source: [IAE 12b]). For a color version of this figure, see www.iste.co.uk/amiard/radioactive.zip*

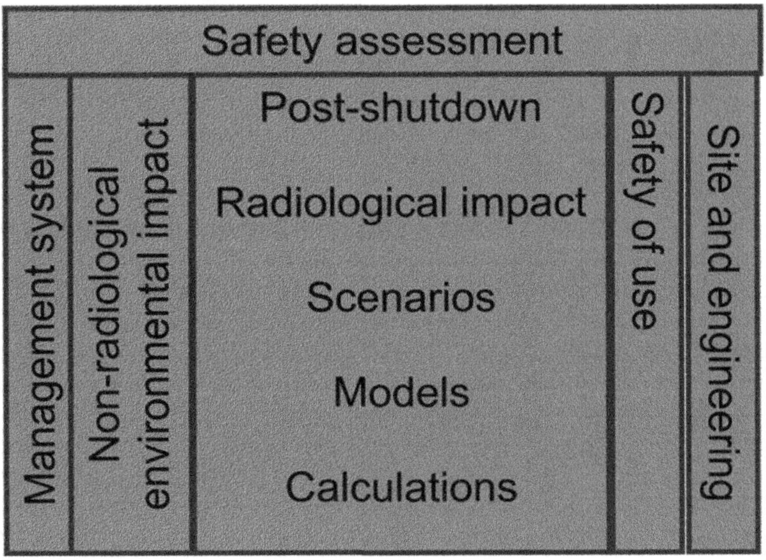

Figure 3.2. *Aspects included in the safety assessment (source: [IAE 12b]). For a color version of this figure, see www.iste.co.uk/amiard/radioactive.zip*

3.3.2. *Recommendations concerning graphite waste*

Graphite has been used quite frequently as a neutron flux moderator in nuclear reactors. It now represents radioactive waste that must be specifically managed. Indeed, the impurities present in new graphite create long-lived radionuclides such as chlorine-36 (302,000 years) by activation, which makes irradiated graphite an LLW-LL waste. The states with the largest quantities of irradiated graphite waste are the United Kingdom (86,000 t), Russia (60,000 t), the United States (55,000 t) and France (23,000 t) [IAE 16a]. The ASN [ASN 20a] estimates that there is a total of 81,000 m^3 of graphite waste.

In France, graphite-based nuclear waste results from the dismantling of plutonium reactors (G1, G2 and G3) at Marcoule, first-generation electronuclear reactors (Chinon A1, A2, A3; Saint-Laurent A1, A2 and Bugey), some experimental reactors of the CEA, and from the reprocessing of irradiated UNGG fuel in the UP2-400 plant at La Hague [IAE 16a].

In the United States, 34 reactors have used graphite as a moderator. These are power reactors (St Vrain, Peach Bottom), experimental reactors (SRE, MSRE), university, research, test and plutonium (Hanford). In Germany, five research reactors are graphite-moderated (MTR, Materials Test Reactor) (MERLIN, RFR-3 and -4, DIDO and AVR) [IAE 16a].

3.3.3. *Radioactive waste management solutions*

The IAEA compares seven options for managing radioactive waste and examines whether or not they are compatible with eight categories of waste (Table 3.11). Preferable solutions (green) mean that there is a perfect match; others are acceptable (light green). On the other hand, some solutions are impossible, either for safety reasons (red) or for technical reasons (purple), and finally some solutions need to be evaluated on certain technical or economic aspects (gray).

A: decay storage; B: surface trench; C: retention dam; D: surface installation; E: intermediate geological repository; F: deep geological repository; G: disposal by drilling; DSRS: disused sealed radioactive source; HLW: high-level waste; ILW: intermediate-level waste; LLW: low-level waste; SNF: NORM: naturally occurring radioactive materials; spent nuclear fuel; VLLW: very low-level waste; VSLW: very short-lived waste

Radioactive waste flows		Final point						
		A	B	C	D	E	F	G
VSLW	Small volume							
	Large volume						NC	NC
VLLW	Small volume							
	Large volume							
LLW	Small volume							
	Large volume							
ILW	Small volume							
	Large volume							
SNF HLW								
DSRS	Small volume							
	Large volume							
	SHARS							
NORM	Small volume							
	Large volume							
Uranium M&M	Small volume							
	Large volume							

Impossible for security reasons	Impossible for technical reasons	Possible but need to evaluate technical or economic aspects	Acceptable	Preferable

Table 3.11. *Example screening matrix for initial identification of feasible disposal alternatives courtesy of Peter Opsial IAEA (source: [IAE 17a]). Legend. For a color version of this table, see www.iste.co.uk/amiard/radioactive.zip*

The IAEA identifies five different cases (cases A–E) of states according to their use of nuclear energy and therefore the types and quantities of waste they generate [IAE 17a].

The typical annual waste flow, in m^3 per GW produced annually, for a state with a complete fuel cycle is 155 m^3 (liquid and solid wasteS) for the upstream part of the cycle (conversion, enrichment, fabrications), 70 m^3 of liquid waste and 260 m^3 of solid waste for the operation of power reactors, and 140 m^3 of liquid waste and 80 m^3 of solid waste for the downstream part of the fuel cycle (reprocessing). The dismantling of the various stages generates nearly 390 m^3 of solid waste per gigawatt produced [IAE 17a].

3.3.4. *Waiting and processing time for nuclear fuel*

In the case of nuclear fuel and the various treatments that can be envisaged for its reuse, the heat released and the intense irradiation require varying latency times before it can be handled (Table 3.12).

Product or service	**Duration of the process**
Natural uranium	2 years before loading
Conversion to UF_6	1.5 years before loading
Enrichment	1 year before loading
UO_2 fuel manufacturing	0.5 year before loading
Fuel storage in the reactor before reprocessing	2 years after loading
Storage of UO_2 or ERU spent fuel before reprocessing	5 years after loading
Storage of MOX spent fuel before reprocessing	5 years after loading
Reprocessing (Pu and U)	1 year
MOX fuel manufacturing	1 year
ERU fuel manufacturing	0.5 years

Table 3.12. *Waiting and processing times for a nuclear fuel cycle for a thermal reactor (source: [IAE 19c])*

3.3.5. *The need for teaching*

As in the case of all activities in the nuclear field, personnel must be competent and informed about the development of regulations, methods and techniques to perform good radioactive waste management. All personnel should therefore receive regular training on this subject. The IAEA [IAE 18b] recommends the best procedures for carrying out relevant training in this respect and how to evaluate the quality of this training. In France, training in radioactive waste management is offered, for example, by the INSTN (*Institut national des sciences et techniques nucléaires*) of the CEA (http://www-instn.cea.fr/formations/formations-continues/liste-des-parcours-de-formation/parcours-gestion-des-dechets.html).

3.4. Some examples of radioactive waste management

Just as each state has its own classification of radioactive waste, so their management varies greatly from one country to another. Attempts to converge both classifications and inventories are being made by the IAEA and the NEA.

3.4.1. *International inventories of radioactive waste*

At the global level, the inventory of the various categories of radioactive waste is provided in Table 3.13. The data feedback from the various states explains the relative age of the values (from 2013 to 2016). The LLW category generally represents the largest volumes.

	European Union (2013)		**United Kingdom (2016)**	**United States (2014)**	**Canada (2013)**
Category	Volume (m^3)	Percentage (%)	Volume (m^3)	Volume (m^3)	Volume (m^3)
VLLW	516,000	15	2,220,000		
LLW	2,453,000	74	1,600,000	$1.51.10^7$	2,499,803
ILW	338,000	10	449,000		67,738
HLW	6,000	0.2	1,500		20,666
Spent fuel	54,300	-		$9.10.10^4$	

Table 3.13. *Quantities of radioactive waste and spent fuel in the European Union and other states on various dates (source: [AND 18c])*

3.4.2. *Surface storage*

Many states store their low- and intermediate-level waste, especially short-lived waste, in centralized centers, and these are above ground (Table 3.14). In contrast, the United States has a multitude of local storage centers generally in the vicinity of power reactors. Thus, in the United States, in 1999, LLW was distributed in nine states and in 36 sites [BLA 01].

Some states have decided to use old, partially decommissioned plants to store LLWR. Such situations have occurred in the United Kingdom (Northwest England) and at sites in the Russian Federation (FSUE RADON, Federal State Unitary Enterprise) [CRO 17].

Deposit, Country	**Opening of the deposit**	**Capacity (m^3)**
Dukivany, Czech Republic	1995	522.10^4
Centre de l'Aube, France	1992	1.10^6
Rokkasho, Japan	1992	4.10^4
SIA Radon, Russia	1961	$1.9.10^5$
El Cabril, Spain	1992	5.10^4
Drigg, United Kingdom	1959	$1.8.10^6$

Table 3.14. *Examples of near-surface disposal facilities (NSDFs) (source: [OJO 14])*

3.4.2.1. *Belgium*

Belgium has decided on surface storage for low- and intermediate-level short-lived waste. Four sites were pre-selected in Dessel and Mol and were approved by the municipal authorities concerned in 2005. In 2006, the government chose to place category A waste in a surface disposal site in Dessel, on the basis of a preliminary project developed with this community.

In 2013, Ondraf (*Organisme National des Déchets Radioactifs et des Matières Fissiles*) then applied to the Federal Nuclear Control Agency (AFCN) for authorization to create a nuclear facility for this storage. The Agency asked Ondraf some 300 questions. In 2019, Ondraf submitted its amended safety dossier to the AFCN, which forwarded it to its scientific council for an opinion. The public consultation took place from April 15, 2020 to June 13, 2020. The authorization could be issued very soon. The first waste would then be stored in the installation in 2024 [CNE 19a].

3.4.2.2. *Japan*

Japan has plans for four types of repositories (surface, subsurface, underground and deep geological) [NII 15]. They are described in Figure 3.3, and the various engineered barriers required are shown in Figure 3.4.

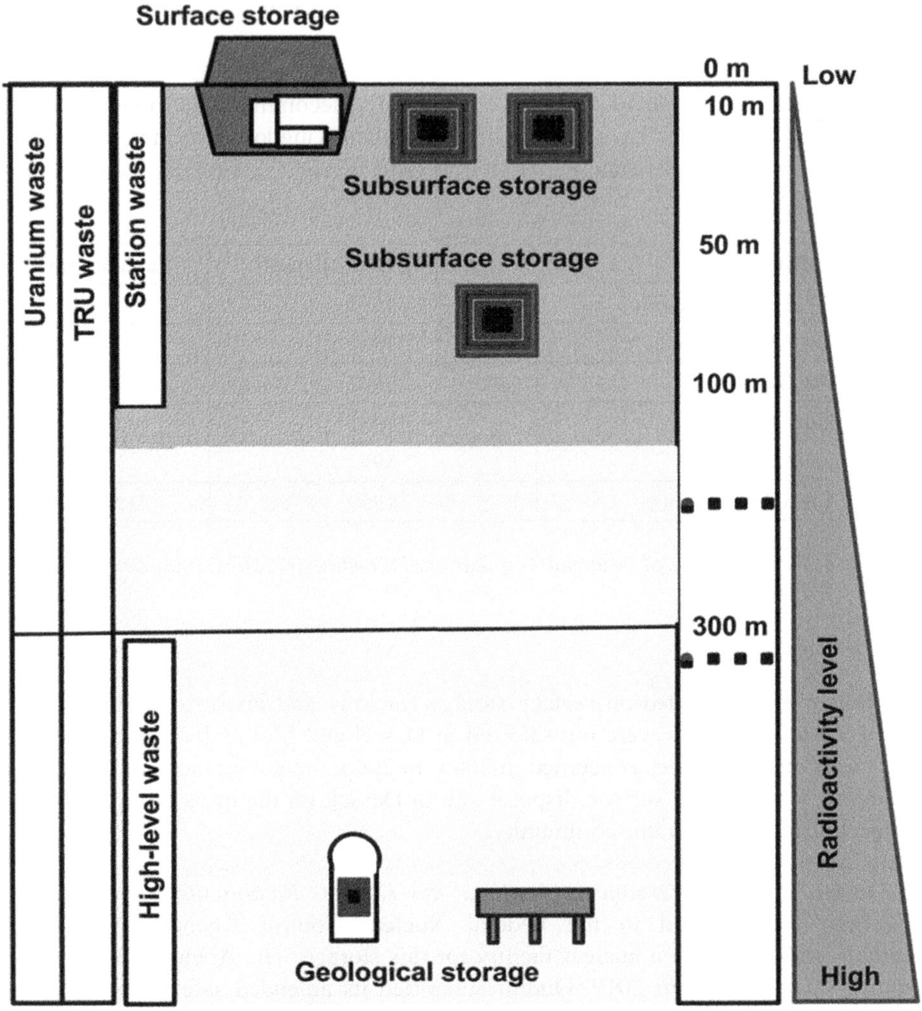

Figure 3.3. *The four Japanese radioactive waste disposal projects (source: [NII 15]). For a color version of this figure, see www.iste.co.uk/amiard/radioactive.zip*

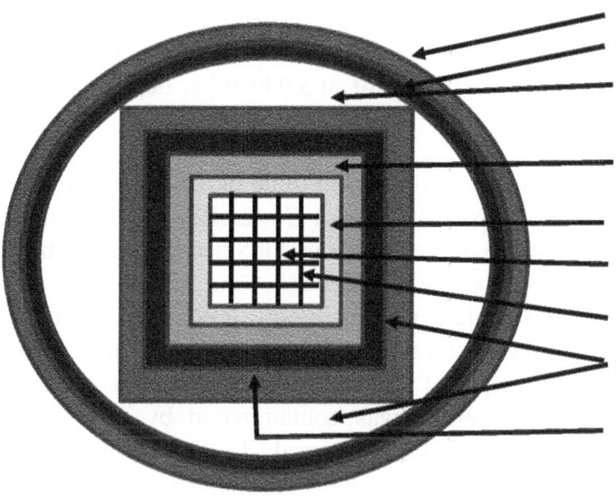

Figure 3.4. *Artificial barriers for subsurface storage of radioactive waste (source: [NII 15]). For a color version of this figure, see www.iste.co.uk/amiard/radioactive.zip*

3.4.3. *Geological disposal of radioactive waste*

Conversely, some countries now have underground facilities for low-level radioactive waste, notably Germany, Sweden and Finland. The choice of host rock differs from country to country. For example, the research and programs developed in Sweden and Finland are focused on storage in granitic rock. Granite is also studied in Korea, Japan, Switzerland and China. On the other hand, clay formations have been the subject of important studies and research for many years in France (Bure in the Meuse/Haute-Marne region), in Belgium (Boom clay) and in Switzerland (Opaline clay). As for Germany, it has opted for salt formations

3.4.3.1. *Germany*

In Germany, the Morsleben repository for low- and intermediate-level short-lived radioactive waste is located in a former rock salt and potash mine in a salt structure in the Aller Valley fault zone in northeast Germany. By the end of the operational phase, a waste volume of about 37,000 m^3 with a total activity of about 450 TBq had been disposed of. Most important for the safety of the storage is the limitation of the inflow of water into the salt structure by the rock cover, which is watertight [RAN 02]. This site was used until 1998.

Several other sites have been explored, such as the salt dome in Gorleden. Another site has been authorized for geological disposal of all German radioactive waste that does not release heat. This is a former iron ore mine in a sedimentary formation, the Konrad mine in the town of Salzgitter near Hanover in southeast Lower Saxony. Some low- and intermediate-level radioactive waste was stored there until 1978. The authorization covers a maximum of 303,000 m^3 at depths of about 1,000 meters.

A final storage site is the Asse II mine, a former salt mine in the Asse mountains of Wolfenbüttel in Lower Saxony. Between 1967 and 1978, LLRW and ILRW was placed in 13 chambers of the mine. An inventory made in 2008 listed 125,787 drums of LLRW stored from 1967 to 1978 in various chambers at the 750 m level for a total activity of $1.8.10^{15}$ Bq and 1,293 containers containing ILRW stored from 1972 to 1977 in chamber 8a at the 511 m level for a total activity of $2.8.10^{15}$ Bq. A deposit of 11 kg of plutonium would be at a depth of 750 m. Numerous ingresses of water were noted as early as 1980 and the brine was contaminated by cesium-137, plutonium and strontium-90. The daily pumping was already 12,000 L in 1988. This finally led the German authorities to decide in 2008 to remove the waste and remediate the mine. Faced with this disastrous situation, the Federal Office for Radiation Protection (BfS) became the owner of the mine and it was decided (law of April 24, 2013) to evacuate all the waste, to recondition it and to find a new site for its final disposal (Dose *et al.* in [IAE 16b]).

3.4.3.2. *Finland*

In Finland and Sweden, storage facilities are currently in operation and their characteristics are that they are accessible from the land, but are located under the sea.

In Finland, two storage facilities have been excavated in granite formations at depths of about 70–100 m to receive operational waste from the Olkiluoto and Loviisa power plants. These facilities, located in the vicinity of these two power plants, have been in operation since 1992 and 1997. The Loviisa storage facility is 90 km east of Helsinki. It receives low- and intermediate-level waste. It is dug into granite at a depth of about 110 m in the Gulf of Finland, from a peninsula of the same name. An extension of this facility is planned to accommodate waste produced during the dismantling of the nuclear power plant located on the same site [STU 17].

The Olkiluoto repository for LLW and ILW is located in bedrock several dozen meters below the surface. It has separate canisters and tunnels for LLW and ILW, and has been designed to collect all the radioactive operational waste produced during the life of the plant. When all the waste has been placed in it, the tunnels and shafts leading to the silos will be filled in and sealed [IRS 19c].

3.4.3.3. *Sweden*

In Sweden, a ship (the "*Sigyn*") has been specially built to transport radioactive shipments from nuclear power plants to the final disposal center for radioactive operational waste with low to intermediate radioactivity and a short half-life. These centers are located near the Forsmark nuclear power plant (called SFR, Repository for short-lived radioactive waste) and near the Oskarshamn nuclear site for spent nuclear fuel (CLAB, Central interim storage facility for spent nuclear fuel) [MEA 07].

The SFR storage facility in Forsmark is located 140 km north of Stockholm. It was commissioned in 1988 and is operated by SKB. It is dedicated to low- and intermediate-level short-lived waste from power plant operations as well as from medical, industrial and research applications. It consists of four 160 m long cavities and a 50 m high silo, dug into the crystalline rock 50–60 m below the level of the Baltic Sea. Access to the facility is from land through two parallel tunnels (access tunnel and construction tunnel) 1 km long near the port of Forsmark. In 2018, the Swedish safety authority gave its approval to the operator's 2014 application to extend the facility [SKB 18], to excavate six new cavities [IRS 19c].

The new vaults of the extension will be deeper, about 120 m below the sea floor. This is where studies have shown that the rock has suitable properties. The new section of the SFR will include five 275 m long rock vaults and one 240 m long vault. When the expansion is complete, the SFR will be three times its current size. The final storage of both operational and decommissioning LLW and ILW will be consolidated on one site and in one facility.

The existing SFR can hold about 63,000 m^3 of operational LLW and ILW and is now more than half full. In the future, there will be almost twice as much decommissioning waste as operating waste. The total amount of decommissioning waste and operating waste is estimated at 180,000 m^3 [SKB 18]. Construction is scheduled to begin in 2023 and operation in 2029 [CNE 19a].

The Swedish Repository for long-lived waste (SFL) is an ultimate disposal facility. It will be needed for the LLW from the Swedish program. Inventory estimation, development of dedicated technology concepts and safety assessments are underway. The required capacity is rather limited as it will concern about 10,000 m^3 of conditioned waste. The SFL should be constructed at a depth of 300–500 m. The site has not yet been chosen, and construction is planned after 2030 [CNE 19a].

3.4.3.4. *Canada*

In Canada, three storage projects have been publicly announced. They are controlled by three operators: Ontario Power Generation (OPG), the *Société de gestion des déchets nucléaires* (SGDN) and the *Laboratoires Nucléaires Canadiens* (LNC).

OPG plans to store low- and intermediate-level long-lived waste near the Bruce Nuclear Generating Station in Kincardine, Ontario. The SGDN is looking for a site to store all of Canada's spent fuel and high-level waste. The LNC is proposing to store up to 1 million m^3 of LLW from research, operation and decommissioning in subsurface storage, primarily at the Chalk River National Laboratory site in Ottawa [CNE 19a].

The most advanced project appears to be that of the OPG. It has submitted a DGR (Deep Geologic Repository) project description for its LLW and ILW in sedimentary rock beneath the Bruce site in Ontario. In 2006, the CCSN published a Regulatory Guide G320, which describes, among other things, the safety case concept to be applied. In 2007, the CCSN launched the Coordinated Assessment and Research Program (*Programme coordonné d'évaluation et de recherche*, PCER) to study sedimentary rock formations at the Bruce site. In 2011, the OPG submitted the Environmental Impact Statement for the DFGP. This was reviewed by the CCSN and the CEC (*Commission d'examen conjoint*) in 2012. In 2015, the CEC submitted its report for the DFGP proposed by the OPG to the Minister of the Environment, with a positive recommendation for the project [CCS 17].

3.4.3.5. *Other countries*

Other countries, such as Korea and Hungary, are considering the use of underground facilities to store their LLW and ILW, both long- and short-lived [IRS 13d].

In Brazil, nuclear waste management remains an unresolved issue. Most of the waste produced in Brazil is classified with low and medium radiation levels. Therefore, the national repository will be near the surface, in accordance with legislation. Considering the concept of multiple barriers for disposal, the radioactive waste is the first barrier. To have a qualified radioactive waste, it must be solid or solidified with an inert material. In order to standardize the disposal process, all radioactive waste will be placed in concrete containers. These containers will be installed in a concrete cell, which is the final engineered barrier in the repository. The operating and monitoring periods of the repository are 60 and 300 years, respectively [TAV 19]. Machado *et al.* [MAC 19] consider that clay materials should be used as a buffer in a repository for LLW and ILW.

3.5. Radioactive waste outside the nuclear fuel cycle

Besides the use of nuclear energy to produce electricity involving all stages of the nuclear fuel cycle, there are a large number of other applications that also generate radioactive waste. The main characteristic of this radioactive waste is that it comes from small producers and is therefore widely dispersed. Their origins are health care institutions, research centers and various industries.

3.5.1. *Hospital and healthcare waste*

Healthcare facilities use two broad categories of radionuclides, radioactive tracers, usually in liquid form, and sealed sources, to fulfil the two applications of diagnostics or therapy.

Radionuclides used as tracers in hospitals have relatively short half-lives. In practice, they are divided into three types: type I with very short physical half-lives (less than 6 days), type II with short half-lives (6–71 days) and type III with long half-lives (more than 71 days) [IAE 00]. In France, the upper limit of VSL waste is 100 days.

In France, type I waste is kept on site for two months and type II for one or two years, so that physical decay has caused the radioactivity to decrease significantly and so that the waste can be directly eliminated into the environment in a normalized manner. For type III waste, ANDRA takes charge of it after sorting it by category in the establishment that employed it [PEY 92]. In France, the discharge limit for liquid effluents is 10 $Bq.L^{-1}$, except for the rooms of patients treated with iodine-131, where this value is raised to 100 $Bq.L^{-1}$ [AND 18c].

In medicine, radioactive tracers are injected into patients, often in large quantities for easy detection. The radionuclides used at more than one GBq are ^{3}H, ^{13}N, ^{14}C, ^{22}Na, ^{35}S, ^{38}K, ^{67}Ga, ^{68}Ga, ^{67}Cu, ^{81m}Kr, ^{99m}Tc, ^{125}I, ^{131}I, ^{113}Sn and ^{153}Sm [IAE 00].

Radioactive waste from medical environment is either liquid or gaseous effluents, or contaminated solid or liquid waste [AND 18c]. The effluents come mainly from the rinsing of equipment and patient excretions (urine and feces). Contaminated waste may be sharp and is also likely to present infectious, chemical or toxic risks.

In France, type III waste represented only a low annual activity of the order of 37–74 GBq of tritium and 7.4–14.8 GBq of carbon-14, for all French hospital services in 1991 [PEY 92]. At the end of 2015, medical waste represented 1% of the entire French radioactive waste volume. Very short-lived waste, 84% from the medical sector, represented a volume of 2,017 m^3 in 2015 [AND 16d].

Sealed sources used in medicine and medical research are very numerous. The IAEA [IAE 00] lists a large number of types for multiple applications. These sources only become waste at the end of their life.

3.5.2. *Industrial and research waste*

As with the medical community, research centers and industry use radioactivity either through radioactive tracers or with sealed sources. The volume generated by this sector in France at the end of 2015 was 3% of all French radioactive waste [AND 16d].

Sealed sources are used in the industry to sterilize food, eradicate insect pests, protect certain heritage items or sterilize medical materials. They can also be used to control and analyze various materials, notably through non-destructive testing by gamma radiography, such as the control of the presence of lead in paint. Sealed sources can be used to check the weight and thickness of paper sheets, or the filling level of tanks or other containers.

Sealed sources for industrial and medical uses, if used properly, do not pose health problems. On the contrary, various accidents have occurred [AMI 19]. On the other hand, at the end of their life, these sources must be rigorously managed to avoid radiation accidents. Their management varies from one country to another. In France, they are managed by ANDRA, which is currently storing them until a final solution is found [AND 18c].

In France, to increase the ionization of the air and the effectiveness of lightning conductors, sources of radium-226 (50 MBq) or americium-241 (20 MBq) have been used. They were manufactured from 1932 to 1983. In total, there are 30,000 lightning rods with a radium-226 source (sometimes mixed with an americium-241 source) and 20,000 lightning rods with an americium-241 source. This has been prohibited since 1987. This type of lightning rod is gradually being dismantled and put in its entirety with the LL-LLW. ANDRA now proposes to consider as waste only the radium or americium pellets, which will limit the volume of waste [AND 18c].

3.6. Conclusions

Historically, very low-level radioactive effluents have been directly released into the environment, into rivers or reservoirs, or into the ocean, or injected into the soil. Solid low-level radioactive waste was immersed in the sea.

Low-level but long-lived mining waste was often stored in the ground. The same is true for various applications of radium in the medical field or in watchmaking. Wastes of military origin such as spent fuel, depleted uranium, plutonium and tritium are stored in many nuclear states awaiting a final solution.

The IAEA very quickly issued common sense rules to ensure that radioactive waste management is carried out under good conditions. Each State is responsible for setting up its own policy.

Solutions for very short-lived waste, based on physical decay, are relatively easy to implement and are fairly well applied. On the other hand, some channels are still not in place. This is particularly the case for historical waste, which frequently needs to be reconditioned in a serious manner.

4

Management of Intermediate- and High-level Nuclear Waste

4.1. Introduction

High-level radioactive waste (HLRW) is certainly the thorniest nuclear issue in the nuclear field. The definition of HLW varies from state to state, the main criteria being mass activity and heat release [NEA 16b].

The approval of the nuclear fuel cycle has become the Achilles heel of nuclear power. After more than 50 years of effort, there are currently no active nuclear waste repositories for spent nuclear fuel from commercial nuclear power plants or for HLW from the reprocessing of spent fuel [EWI 16].

The Joint Convention on the Safety of Spent Fuel Management and on the Safety of Radioactive Waste was adopted on September 5, 1997 at a diplomatic conference convened by the IAEA in Vienna. It was opened for signature on September 29, 1997, and entered into force on June 18, 2001 [IAE 17b]. As of December 31, 2018, 80 contracting parties had ratified this convention [TON 06].

Most countries with nuclear power plants are studying techniques for the final disposal of high-level waste in deep geological formations. Thus, out of the 28 states with power and/or research reactors, and therefore HLW, 22 states have planned a deep geological repository (DGR). Some sites have already been selected, such as Oikuluodo in Finland and Forsmark in Sweden. The American project of Yucca Mountain has been suspended [NEA 16b].

Similarly, several countries are studying or building industrial-scale reprocessing and vitrification facilities. Many have instituted special funding for nuclear waste management and disposal programs, and for nuclear power plant decommissioning operations [ZHU 89].

For final storage of high-level waste or spent fuel, the choice of deep storage leads to studies to find a geological site. States are building or planning to build an underground laboratory for future construction and/or research. The packaging adopted for the storage is often vitrified waste containers. On the other hand, the geological type of the site varies between clay, granite, salt and shale.

To manage these administratively, technically and economically burdensome programs, states have created a specialized structure (Table 4.1).

Country	**Acronym**	**Organization**	**Creation**
Germany	BFS	*Bundesamt für Strahlenschutz*	1989
Belgium	ONDRAF	*Organisme national des déchets radioactifs et des matières fissiles enrichies*	1980
Canada	NWMO	Nuclear Waste Management Organization	2002
China	EEE	Everclean Environment Engineering Corp.	1995
Spain	ENRESA	*Empresa Nacional de Residuos Radioactivos, S.A.*	1984
United States	OCRWM	*Office of Civilian Radioactive Waste Management*	1982
Finland	POSIVA	*Posiva Oy*	1995
France	ANDRA	*Agence national pour la gestion des déchets radioactifs*	1991
Japan	NUMO	Nuclear Waste Management Organization of Japan	2000
Netherlands	COVRA	Central Organization for Radioactive Waste	1982
United Kingdom	NOA	Nuclear Decommissioning Authority	2005
Sweden	SKB	*Svensk Kämbränsiehantering AB*	1970
Switzerland	NAGRA	*Nationale Genossenschaft für die Lageung radioaktiver Abfalle*	1972

Table 4.1. *List of waste agencies in various states (source: [PNG 10])*

Risk assessment for radioactive waste disposal facilities is a major challenge in the nuclear field. Indeed, it is necessary to ensure the exhaustiveness of all the scenarios selected to assess the safety of the repository. In particular, it is necessary to answer several questions. First of all, it is necessary to clarify why the scenarios generally focus on a restricted set of functionalities, events and processes. Yet, there is still no consensus on the interpretation of exhaustiveness to guide scenario generation. Finally, there is a need for sound approaches to deal with epistemic uncertainties [TOS 17].

4.2. International recommendations of the IAEA and NEA

The IAEA and NEA have organized many expert meetings, conferences and congresses on the subject of high-level radioactive waste, publishing numerous documents and making many recommendations.

There is a global consensus in the international community that geological repositories can provide the necessary long-term safety and security to isolate long-lived radioactive waste from the human environment over long periods of time [NEA 17c].

The IAEA has launched a project to examine the safety of deep geological repositories for high-level waste, GEOSAF (IAEA, International Project on Demonstrating the Safety of Geological Disposal). This project was carried out in three phases. The first phase, GEOSAF I (2008–2011), surveyed the work done around the world on this subject [IAE 11b]. During the second phase, GEOSAF II (2012–2015), a bibliographical review was initiated [DEP 15] and recommendations concerning the constitution of the safety file were made before the final closure of the repository [IAE 15]. The third phase, GEOSAF III (2017–), is in progress [IAE 17c].

4.2.1. *Spent fuel management*

Spent (or irradiated) fuel is considered as radioactive waste or nuclear material depending on whether a country has opted for the open or closed nuclear fuel cycle. Spent fuel management varies from country to country. Some have centralized management of spent fuel, such as Dessel and Tihange in Belgium, Dukovany in the Czech Republic, La Hague in France, Ahaus and Gorleben in Germany, Mitsu in Japan, ISFSF in Hungary, COVRA in the Netherlands, IFE in Norway, Jaslovské Bohunice in Slovakia, Mountain Mining Combine (MCC) in Russia, CLAB in Sweden, Zentrales Zwischenlager (ZZL) in Switzerland and Sellafield in the United Kingdom [NEA 16b]. The other countries have a dispersed management in the vicinity of nuclear reactors.

The IAEA's recommendations concerning this subject are numerous. For the storage of spent nuclear fuel, the IAEA has published a specific safety guide

[IAE 12a]. The same is true for the preparation of the safety file and the safety assessment for the disposal of radioactive waste; the specific safety guide was published in 2012 [IAE 12b].

The IAEA regularly organizes conferences on the issues raised by spent fuel. Safety of disposal remains a concern, particularly with respect to the availability of geological disposal facilities for high-level waste and spent fuel. This agency has recently highlighted the progress of some countries in this regard [IAE 19c].

The IAEA has also proposed a methodology for the Systematic Assessment of the Regulatory Competence Needs (SARCoN) for regulatory bodies of radiation facilities and activities [IAE 19d].

4.2.2. *Management of radioactive waste resulting from a nuclear accident*

The management of waste following a nuclear accident, and in particular of high-level waste, remains unknown. Our knowledge is based only on a few concrete examples such as Chernobyl, Fukushima, Windscale and Goiânia. Past experience also includes the restoration of mining operations (Wismut), industrial sites that have experienced many accidents with their waste (Hanford) or former atomic test sites (Maralinga) [IAE 17b].

The first piece of information that stands out is the enormous amount of radioactive waste. Thus, in Chernobyl, the waste represents 2,800,000 m^3 with an activity greater than $8.5.10^{15}$ Bq; in Fukushima, it is 16 to 22 million m^3; and in Goiâna 3,400 m^3, it is 46 TBq.

The Wismut Uranium Ore Residues Remediation involves 325 million m^3 of waste rock and 160 million m^3 of slurry tailings from the mining operation. The treatment involves 865,000 m^3 of debris and concrete, 14.5 million m^3 of contaminated soil and waste rock and 200,000 tons of scrap metal from the clean-up and wastewater from the treatments. All of this waste is located in areas that were densely populated [IAE 17b].

The Hanford Site Legacy Production Facility Remediation represents 200,000 m^3 of high-level liquid radioactive waste and 710,000 m^3 of solid radioactive waste [PAU 13].

The Maralinga Legacy Nuclear Weapon Test Site Remediation represents 263,000 m^3 of contaminated soil that was disposed of in a large trench near the surface [DEP 03].

The IAEA proposes numerous documents for managing large volumes of waste from a nuclear accident for environmental remediation, pre-storage management, storage and decommissioning [IAE 17b].

The various stages in the management of radioactive waste generated by a nuclear accident are shown in Figure 4.1. This diagram only works for LLW and ILW, but not for corium, for which no solution is yet available.

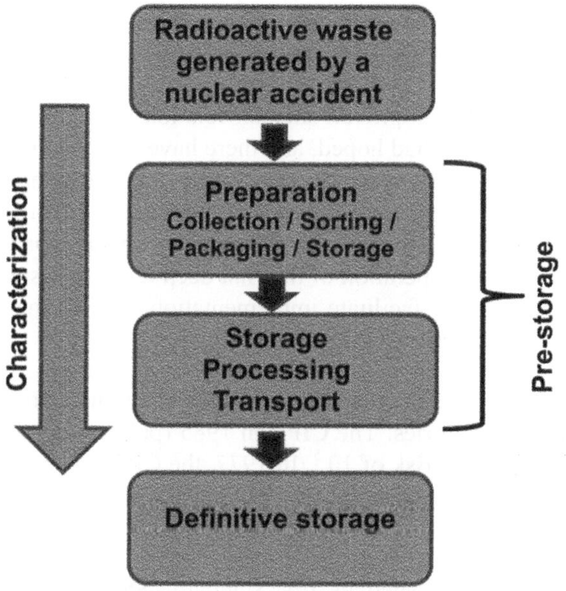

Figure 4.1. *Various stages in the management of low- and intermediate-level radioactive waste generated by a nuclear accident (source: [IAE 17b]). For a color version of this figure, see www.iste.co.uk/amiard/radioactive.zip*

4.2.3. *Final repositories in deep geological layers*

The solution of disposal in deep geological layers has been chosen for many toxic wastes [BUS 18]. For the management of long-lived radioactive waste, extensive debates and discussions have taken place, and the path generally chosen is to place long-lived waste in a geological repository, i.e. an underground repository built at great depth in a carefully chosen rock formation. Other solutions have been considered, such as burying the waste in the seabed and in ultra-deep boreholes or launching it into space. However, these have proved too costly or too risky, or impractical for political or legal reasons.

Several original options such as long-term surface storage and separation and transmutation have been considered. However, storage is only temporary and the problem will arise again in the future (perhaps with technological advances). Separation and transmutation are not real alternatives either. In the best of cases, these techniques make it possible to reduce the volume or to modify the composition of the radionuclides in the waste that must be placed in deep storage.

In the case of programs involving relatively homogeneous geological formations, the engineered barriers will essentially be used to ensure safety during the repository's operation phase, and the natural barriers will take over from there.

Technical advances and improved societal relationships have taken longer than waste management agencies had hoped, and there have been significant delays in the development of deep repositories. However, one repository has recently begun operation, and several deep repositories are approaching commissioning time, although most such facilities cannot be expected to operate for many years. The NEA hopes that successful operation of the first deep repositories will restore public confidence and, in so doing, facilitate implementation of the remaining repositories [NEA 00]. This mistrust persists (see section 3.1).

Many international organizations have issued recommendations to ensure the safety of geological repositories. The CIPR in 1985 (publication 46) set a dose limit of 1 $mSv.yr^{-1}$ and an annual risk of 10^{-5}. In 1977, the CIPR lowered the dose limit to 0.3 $mSv.yr^{-1}$. In 1981, it provided safety criteria for storage at 1,000 and 10,000 years.

For its part, the IAEA issued its first standard (SS-99) in 1989, followed by SS 111-F in 1995, WS-R-4 and SF-1 in 2006, establishing the foundations of safety.

The NEA created the Radioactive Waste Management Committee (RWMC) in 1975 and then held the Cordoba Workshop in 1997 to address the safety issues of long-term storage and to initiate a dialogue between regulators and operators.

In 2000, the NEA produced a report evaluating deep geological repositories. It considers that three safety criteria are essential to ensure the disposal of radioactive waste. These are the risk and dose that must be controlled, the protection of the environment that must be guaranteed and the safety of the repository that must be maintained over a wide time scale. The NEA concludes that strict compliance with quantitative limit values is no longer sufficient because of the time periods involved. Qualitative arguments must be taken into account, such as site selection, good design and engineering, an optimization process, the use of best available techniques and the implementation of adequate management principles. In addition to the protection of human life, the protection of the environment must also be explicitly required. The decision-making process must be carried out in stages, with the possibility of reversing the individual stages [AEN 10c].

The IAEA has compared the disadvantages and advantages of various methods of managing long-lived radioactive waste. Experience shows that radioactive waste storage is safe for several decades if active monitoring and maintenance are provided. In contrast, geological disposal promises long-term safety without monitoring and maintenance. While maintenance is easier above ground than underground, institutional controls cannot be maintained during the period when the waste remains hazardous. Similarly, material retrieval is easier from above-ground facilities than from underground facilities, but geological disposal can be developed in stages to maintain the possibility of retrieval for a long period. One of the essential problems that remains to be solved is the transmission of information about the radioactive repository to future generations [IAE 03].

4.2.4. *Site selection criteria*

Some geological formations can be described as highly isolated sites, either because of the absence of water as a means of transporting radionuclides (e.g. salt domes)[1] or because of the extremely low permeability of the rock matrix and the absence of fractures (e.g. layered clay or argillite), which ensure slow, diffusive transport over reasonably long distances (greater than 50 m) to the near-surface environment.

Finally, it should be noted that mature storage programs in countries such as Sweden, Finland and the United States, which are currently moving into a licensing phase, have seen their designs evolve. This evolution is an iterative, adaptive and optimization process driven and guided by a number of factors, including collection of new site characterization data, revision of regulatory standards, adaptation of designs to pre-shutdown and post-shutdown issues, ongoing engineering surveillance and regulatory review regarding the likelihood and potential consequences of natural event scenarios [APT 17].

For some states with no suitable geological structure and/or low volumes of high-level waste, the solution is to participate in a Multinational Repository (MNR). Indeed, a DGR (Deep Geological Repository) with a capacity of 10,000 metric tons can cost only a little more than another one needed to dispose of 5,000 t. This is the

1 In the case of storage in salt, the arrival of water causes a catastrophe because, on the one hand, this rock has no resistance to water and, on the other hand, the water produces a very aggressive brine for the steel of the containers.

IFNEC (International Framework for Nuclear Energy Cooperation's) project (Zagar *et al.*, in [IAE 20]).

4.2.5. *Temporal evolution of a deep geological repository*

A deep geological repository evolves over time in four stages (Figure 4.2). The first stage is the operational stage during which the packages are placed in the repository, followed by the thermal stage when the packages cool down. During these first two stages, the engineered containments play their role perfectly and ensure water tightness (C1) and limit the entry of water (C2). The third stage is the isolation stage that delays the release (R) thanks to the physical containment, which increases resistance to leaching (R1). The last stage is the geological stage that limits diffusion and increases retention (R2), dilution and dispersion (D), as well as public access (L) through deliberate or inadvertent intrusion [AEN 04a].

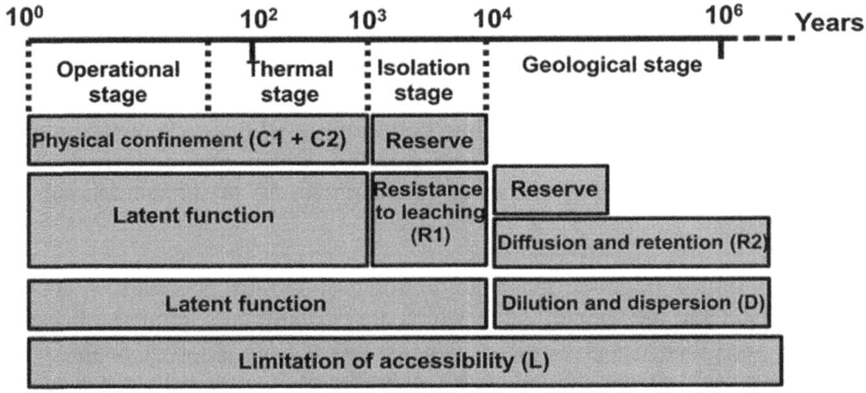

Figure 4.2. *Four stages of the normal evolution of the disposal system proposed by ONDRAF/NIRAS [OND 01] for high-level waste and the corresponding long-term safety functions (source: [DEP 02, AEN 04b]). For a color version of this figure, see www.iste.co.uk/amiard/radioactive.zip*

4.2.6. *Underground laboratory*

With the concept of deep geological storage, operators felt the need to have structures that simulate the future storage in real size. Underground research laboratories (URLs) will gradually see their role refined, their number increased throughout the world and the scientific problems carried out in these structures become more complex. The establishment of an underground research laboratory can be considered as a key element in the success of the repository program [AEN 01].

4.2.6.1. *Roles of underground laboratories*

The primary role of underground laboratories is to test the various possible disposal sites. Thus, in the underground research laboratories of Belgium, Canada, the United States, India, the Federal Republic of Germany, Sweden and Switzerland, scientists and engineers study the properties of possible sites in salt, clay and crystalline rock formations [ZHU 89]. These laboratories provide essential, and at times vital, technical information and confidence for the selection of the storage site, the design of the repository, the constitution of the associated technical basis and the evaluation of safety.

Another advantage of the underground laboratories became apparent quite quickly, and that is the fact that international collaboration developed around them, attracting researchers of various nationalities to the various laboratories.

Underground laboratories can be either entirely dedicated to research and built on sites where no waste will be stored or, on the contrary, designed to study a particular site, in which case the scientific studies and other activities carried out there will be conceived as a preliminary step to the construction and operation of the repository. They provide an excellent opportunity to integrate several disciplines (geology, hydrology and engineering), to build technical teams and to accumulate invaluable hands-on experience in preparation for repository construction [AEN 01].

Over time, the work carried out in underground laboratories has evolved. In the early laboratories, priority was given to the development of test equipment and methodologies and to the conduct of experiments to further our understanding of key mechanisms as well as to the conduct of technical feasibility studies and the collection of fundamental geological data. Today, efforts are more focused on the adaptation and optimization of materials and techniques. In addition, full-scale demonstration experiments on engineered barrier systems are also more frequent [AEN 01].

These underground laboratories will have important roles to play both during repository operation and after closure. The value of underground laboratories to public trust can be considerable. They offer unique opportunities to demonstrate the repository concept and the technical feasibility of the repository construction program, and thus to convince the public that the program is soundly based and conducted responsibly by trustworthy waste managers [AEN 01].

Finally, URLs also make it possible to envisage normal or degraded operation scenarios for the equipment, to test incident and accident situations and to validate the remediation devices [DEL 14].

4.2.6.2. *Different underground laboratories in the world*

The first international underground laboratory was built in the Stripa mine in Sweden and put into operation in 1978 [ZHU 89]. In 1997, 16 underground laboratories existed in the world, in Germany, Belgium, Canada, the United States, Finland, France, Japan, Sweden and Switzerland, notably in salt domes, granite or clay [KIC 97, AEN 00]. Since then the number set up has increased, some having ceased to function. For their part, Ojovan and Lee [OJO 14] list 20 experimental underground installations. The IRSN [IRS 13d] listed 12 in 2013 with two objectives, either to develop knowledge and validate methods and technologies or to characterize a specific site in order to evaluate the feasibility of a storage facility.

Methodological laboratories meeting the first objective have been created in granite in Canada (URL, Underground Research Laboratory), in Whiteshell (now being dismantled), in Sweden (Äspö laboratory), in Switzerland (Grimsel laboratory) and more recently in Korea (*KURT*, Kaeri Underground Research Tunnel) and in Japan (Tono Mizunami URL). The same type exists in clay formations in Belgium (Mol), Switzerland (Mont-Terri) and Japan (Horonobe URL). The Tournemire experimental station operated by IRSN falls into this category.

A second category site characterization and qualification laboratory exists in the United States (Yucca Mountain in the tuff). Another is under construction in Finland (Onkalo on the island of Olkiluoto). ANDRA's underground research laboratory at Bure falls into this second category of research facility [IRS 13d].

The NEA census [NEA 13] is more complete since, for the OECD countries alone, 27 underground laboratories are or have been in operation. Among these, 12 laboratories are based on pre-existing underground excavations. These are Asse Mine (Germany), Tono Mine (Japan with France and Switzerland), Kamaish Mine (Japan with Switzerland), Stripa Mine (Sweden with the international cooperation of seven countries), Grimsel Test Site (Switzerland with the cooperation of 12 countries), Mont Terri (Switzerland with the cooperation of eight countries), Olkiluoto Research Tunnel (Finland with Sweden), Climax (United States), G-tunnel (United States), Amélie (France), Fanay-Augères (France) and Tournemine (France with Canada and IAEA). Seven were specifically designed in NEA member countries. These are HADES-URF (Belgium in cooperation with six countries), AECL Underground Research Laboratory (Canada in cooperation with five countries), Mizunami Underground Research Laboratory (Japan in cooperation with three countries), Honorable Underground Research Laboratory (Japan with France and Switzerland), Aspö Hard Rock Laboratory (Sweden in cooperation with nine countries), Busted Butte (United States) and Korea Underground Research Tunnel

(South Korea). Eight laboratories are specific to the site envisaged as a deep geological repository. They are ONKALO (Finland), Gorteben (Germany), Konrad (Germany), Morsleben (Germany), Pécs (Hungary), Waste Isolation Pilot Plant (the United States in cooperation with six countries), Exploratory Studies Facility (the United States) and Bure (France in cooperation with Switzerland, Germany and Japan) [NEA 13].

4.2.6.3. *Scientific problems dealt with in underground laboratories*

Certain types of information and experience needed to characterize, construct and operate a geological repository can only be obtained by accessing the underground environment. Similarly, confidence in facility design, host rock suitability and engineering feasibility can only be gained through underground verification. All of these factors are important in constructing the safety file [NEA 01].

Underground laboratories are used to answer four scientific questions related to the safety of high-level and long-lived radioactive waste in the case of deep geological disposal. These four scientific problems are (a) the effects of heat on engineered barriers and the geological environment; (b) geochemical characterization of pore water in clay rocks; (c) radionuclide diffusion and retention and (d) full-scale sealing of a repository.

Delay *et al.* [DEL 14] summarize the results obtained from experiments conducted in five underground research laboratories, three of which were in clay formations (Mol in Belgium, Centre de Meuse/Haute-Marne in France and Laboratoire de roches du Mont Terri in Switzerland) and two in granitic rocks (Aspö Hard Rock Laboratory in Sweden and Grimsel Test Site in Switzerland).

Among the specific topics for which the underground laboratories have expanded the knowledge base, one of the most obvious is the invaluable experience that has been gained under actual site conditions for the excavation, and construction and operation of a waste repository [BLE 17].

Underground laboratories may have important future roles during operations and after closure of the deep geological repository site. If waste retrieval were to become a goal, then the laboratory could be used as a test bed for the development of methods, equipment and experiments [NEA 01].

The value of URLs in increasing public confidence can be considerable [NEA 01]. Indeed, URLs provide unique opportunities to demonstrate to the interested general public the scientific and technical work conducted to evaluate the safety of a repository and thus contribute significantly to increasing public confidence in deep geological repositories [DEL 14].

4.2.7. Retrievability and recovery

Faced with the reluctance, or even hostility, of local populations to the introduction of a high-level radioactive waste repository in their vicinity, the authorities of certain states have put forward the hypothesis that the waste can be recovered for a certain period. The countries that have integrated the principle of retrievability into their programs mention three main reasons. The first is that it demonstrates an attitude of humility and openness to the future. The second is that it provides additional guarantees of safety, and finally it responds to the general public's desire not to be trapped in an irreversible situation from the moment the waste is placed in storage [AEN 12b].

The implementation of a policy of reversibility and retrievability allows future generations to have all options at their disposal. However, this raises two questions: how to preserve these options? And for how long is it reasonable and desirable to preserve them? [AEN 12b].

This reversibility option demonstrates that the authorities are proceeding along this path in successive stages, with the necessary caution and flexibility, and that there will be opportunities for revisions based on technical considerations and the public interest. However, even though future generations are given the choice between several management options, the problem of radioactive waste remains the primary responsibility of the present generation.

The introduction of reversibility in a gradual process of repository development, and in particular of measures designed to promote waste retrievability, makes it possible to take advantage of advances in scientific knowledge and technology and to adapt to changes in national policies, regulations and societal positions [AEN 02a].

However, some repository concepts are more amenable to technically and economically sound recovery, provided that backfilling of access routes can be delayed for a long time after the waste has been placed.

There are no circumstances that would justify emergency waste retrieval. Therefore, authorities will always have sufficient time to implement an orderly program and to construct all necessary storage facilities before exploring alternative disposal options. During the waste placement phase and prior to backfilling, recovery will be carried out with the same equipment in a reverse process. After backfilling, special techniques may be required. Indeed, the high temperature and radiation conditions that will persist around the spent fuel and HLW packages will require specific excavation operations [AEN 02a].

Each state shall provide regulatory principles on retrievability and maintain a sufficient level of technical capability at each stage. Recovery methods shall be defined, including for cases of recovery under foreseeable conditions of package failure or accident [AEN 02a].

4.2.8. *Safety file*

All national radioactive waste management authorities now recognize that a strong safety file is essential for the development of radioactive waste disposal facilities [NEA 17b].

Reflections on the composition of a safety file for the operational and post-closure phases of geological repositories for radioactive waste, ranging from LLW to HLW and spent fuel waste, have been conducted by the Nuclear Energy Agency, the European Commission and the International Atomic Energy Agency. Significant differences in focus exist between the three organizations [NEA 17b].

The NEA [AEN 04a] defines the purpose and context of the safety file as follows. The safety file is a specific step in the planning and implementation of the repository system. The safety file includes the implementation and design strategy, the management strategy and the assessment strategy. The basis of the assessment is the system design; scientific and technical information and interpretations; and methods, models, computer codes and databases.

The evidence and analysis and arguments must meet several assurances. These include the intrinsic quality of the site and design, compliance with dose and risk criteria (including reliability of analyses) and the use of safety indicators complementary to dose and/or risk. They also include the ability to meet the uncertainty management strategy and the resolution of outstanding issues. Finally, the demonstration of the robustness of geological disposal as a waste management solution.

The safety assessment of a geological repository is based on safety indicators and reference values or “benchmarks” (Figure 4.3).

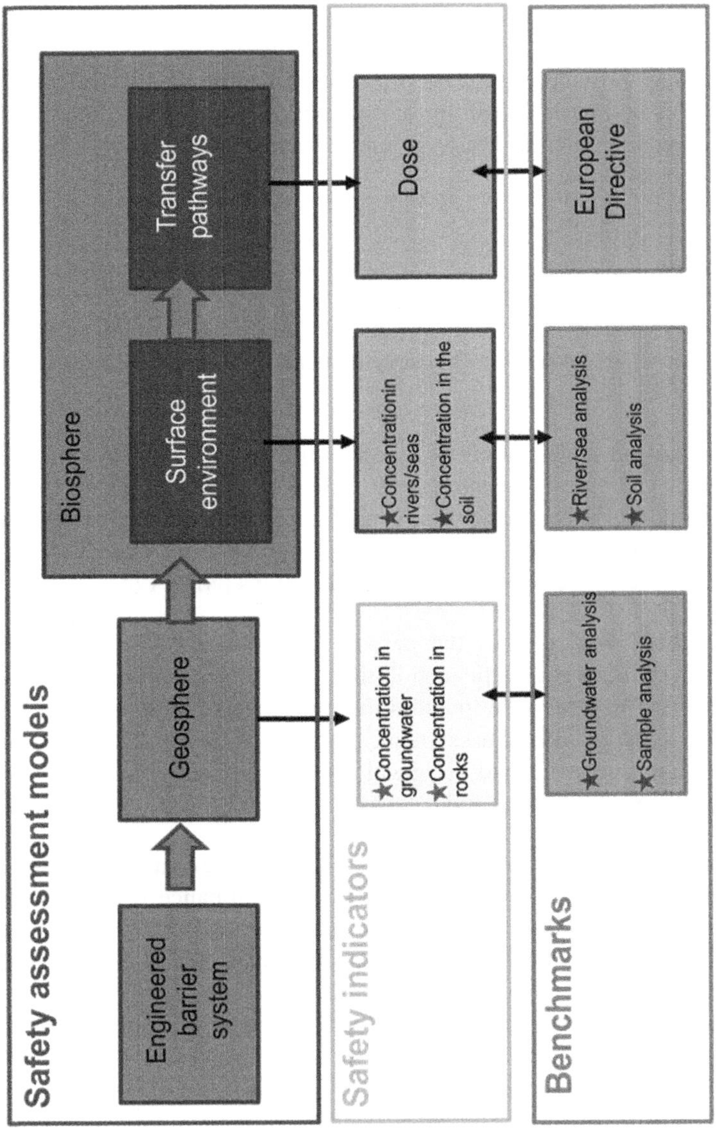

Figure 4.3. *Complementary safety assessment models and corresponding reference values or "standards" used in the safety assessment (source: [UME 02]; [AEN 04b]). For a color version of this figure, see www.iste.co.uk/amiard/radioactive.zip*

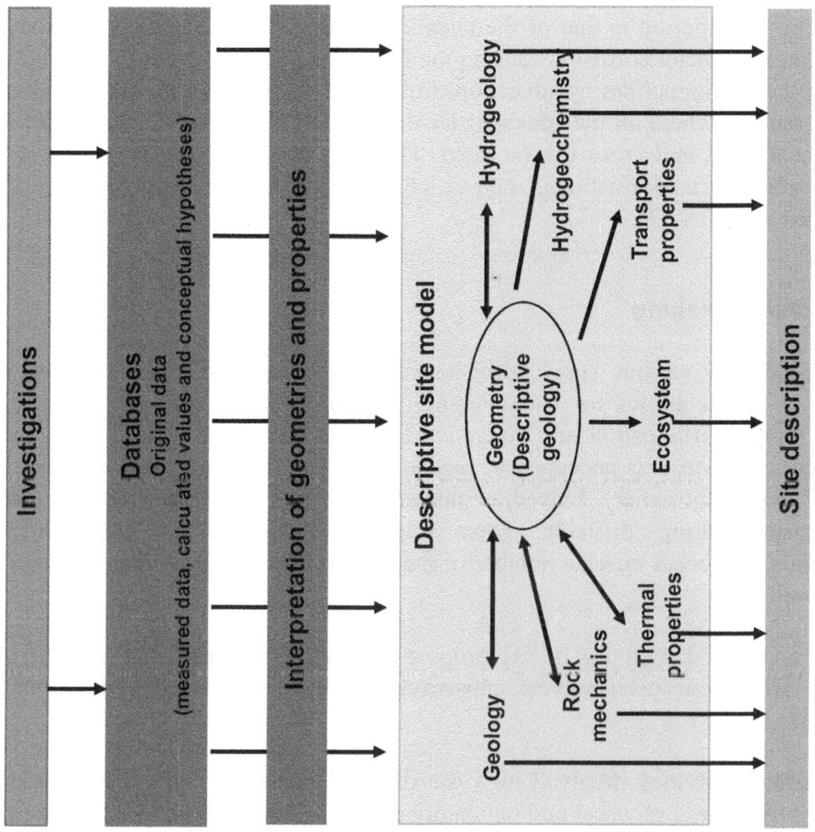

Figure 4.4. *Illustration of how the integrated site description called the SKB and Posiva Site Description Model is performed (source: [AEN 10b]). For a color version of this figure, see www.iste.co.uk/amiard/radioactive.zip*

The description of the storage site must be accurate and complete. Figure 4.4 shows the solution chosen by SKB and Posiva.

The British agency Nirex subdivided the safety assessment into five time periods. The first period corresponds to the installation of the waste containers, where monitoring is mainly concerned with the corrosion of the steel of the containers. The second period corresponds to the evolution of the physical and chemical barriers, where monitoring is carried out using control clones and of the formation of complexes with organic compounds; the emissions are essentially gaseous. The third period is that of the chemical barrier with a completely anoxic environment, significant corrosion causing the failure of a large number of packages and a release of radionuclides by advection-diffusion. Period 4 is the so-called stable geological barrier, where all the packages have failed and the radionuclides migrate from the near field and cross the far field. The fifth period is where the system reacts to external transformations, such as climatic changes and hydrogeological transformations [NEA 02].

4.2.9. *Decision-making*

Increasingly, in various states, decision-making is done in stages. Its main advantage is that it allows for backtracking. In addition, this stepwise approach facilitates public participation and collective learning. However, in the long-term management of waste, it is necessary to reconcile the contradictory requirements of safety and control by society. Indeed, certain social values are in contradiction and make decision-making difficult. Thus, the introduction of a step-by-step decision-making process raises a number of methodological problems that will have to be resolved.

Quinn and Rohrbaugh [QUI 83] propose a theory of conflicting values of a decision from four perspectives: consensual, political, empirical and rational (Figure 4.5).

The various activities involved in waste disposal, including safety, economic, political, geographical, physical and regulatory elements and the decisions that result from them, can extend over a period of 60 years in Sweden, and this in the case of a process without any sudden changes or deviations [PAP 01].

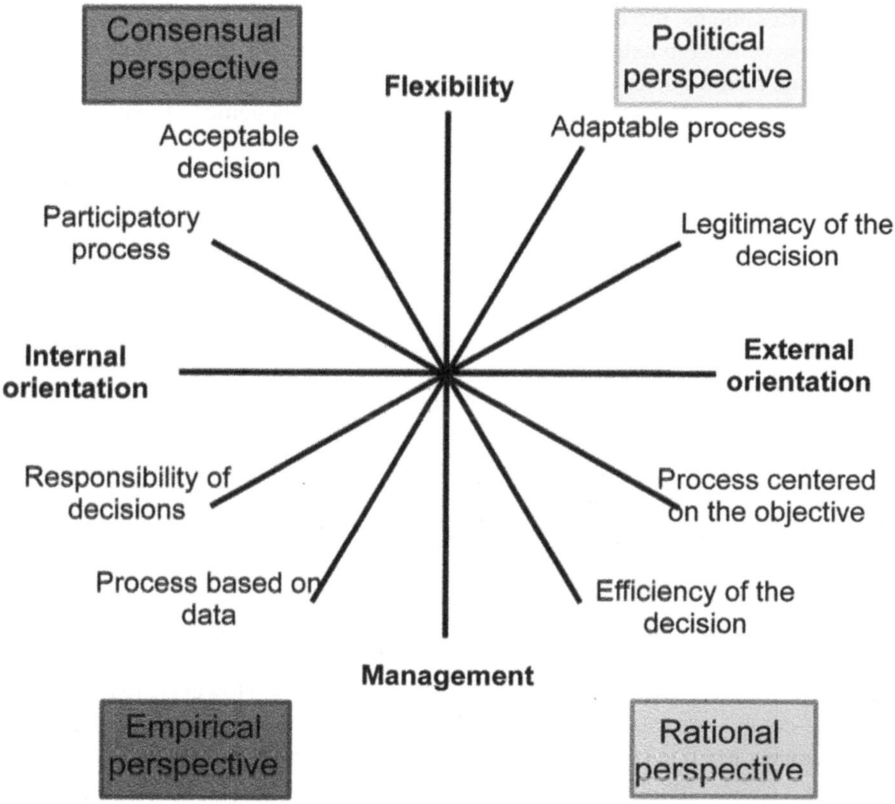

Figure 4.5. *Diagram for the theory of contradictory values (source: [QUI 83]). For a color version of this figure, see www.iste.co.uk/amiard/radioactive.zip*

4.2.10. *Long-term evolution and post-closure monitoring*

The limits of predictability for the various aspects of the geological disposal system are quite different. This limit is only a few decades for the radiological exposure modes, a few hundred years for the surface environmental processes, a few tens of thousands of years for the hydrogeological system and for the performance of the engineered barriers and about 1 million years for the host rock [AEN 04a].

Assessing the stability of the geological environment for nuclear waste repositories requires a long-term (100,000 to 1,000,000 years) understanding of tectonic and climatic processes. This is significantly longer than comparable stability assessments for other critical facilities such as nuclear power plants and large dams, which typically have lifetimes of less than 100 years. Thus, there is a need to understand current stability (frequency and size of earthquakes, proximity and effects of volcanic activity, effects of anthropogenic climate change), cumulative effects over long periods of time (such as the gradual effects of slow uplift and erosion) and how tectonic and climatic processes may change over time (such as the formation of new fault zones or volcanoes and the impact of ice ages) [CLA 17].

4.3. High-level radioactive waste management and the public

Public perception of nuclear energy was negative following the atomic bombings in Hiroshima and Nagasaki. This strong distrust has been maintained with regard to civilian applications of nuclear energy, with the exception of the medical field. It is particularly strong and worldwide with regard to radioactive waste. Waste has a negative aspect that is reinforced by its radioactive nature. Public opposition to any establishment related to the management of radioactive waste is quite systematic. This has been seen for the creation of underground laboratories, for various types of storage and of course for deep geological storage. This phenomenon is reinforced by the various nuclear accidents, all the more so when they take place in a waste repository, such as the WIPP accident in the United States. However, reactions vary from one country to another. The national authorities must take this state of mind into account, provide reliable and serious information to the public, provide attractive accompanying measures, involve the public as much as possible in decision-making at all levels and innovate to maintain complete information for future generations.

4.3.1. *Public perception of the geological repository project*

A comparison of radioactive waste management policies can be made between three countries: Belgium, Canada and France. The organizations in charge of this issue are, respectively, ONDRAF (*Organisme National des Déchets Radioactifs et des Matières Fissiles*), the NWMO (Nuclear Waste Management Organisation) and ANDRA (*Agence Nationale des Déchets Radioactifs*).

From the 1950s to the 1980s, the three states had very closed socio-technical programs, reserved for a “nuclear elite”. The choice to bury the waste was made in small groups, and political decisions were mainly based on expert assessments generally close to the nuclear industry (mainly research conducted by the country’s nuclear energy research centers), which were based on exclusively techno-scientific and economic considerations. Then, the three countries faced a series of local protests that put the existing practice regime under pressure and forced the government to rethink its practices. In all three cases, this tension arose during the selection of the storage site, with operational studies in the field. For Belgium, it was a project for the storage of LLRW, and for the other two countries, a deep geological repository. In all three cases, therefore, it was an external contingent element that pushed the government, overwhelmed by the crisis, to change its practices. The states were obliged to recognize the socio-technical character and it became necessary to broaden the knowledge and the number of criteria to be taken into consideration [PAR 18]. This change in policy was very beneficial for Canada (see section 4.5.3.2), more uncertain for France (see section 4.5.1.2) and without any notable effect for Belgium (see section 4.5.2.3).

The deep geological disposal center for military radioactive waste (WIPP, Waste Isolation Pilot Plant) located in New Mexico (the United States) is designed to receive 176,000 m^3 of so-called transuranic waste (containing, in particular, americium and plutonium) in cavities dug in salt at a depth of about 660 meters. This center suffered two serious accidents in February 2014 with a fire in a truck in the northern zone [AIB 14] and the explosion of a radioactive drum with the release of radioactivity in the southern zone [AIB 15]. The first accident (fire) was attributable to a series of negligent actions on the part of the DOE and the subcontracting organizations, notably concerning the maintenance of the truck, the inadequacy of emergency procedures in the event of fire, the inadequacy of extinguishing equipment in the underground installation and the training of personnel. The second accident was more serious, since about $3.7.10^7$ Bq of americium-241 and plutonium-239 and 240 were released outside the installation for about 15 hours. Twenty-two workers were exposed to americium-241 at a maximum individual dose of 100 µSv. This accident was due to improper packaging of a batch of drums from Los Alamos National Laboratory. The result was a temporary closure of the center, a partial decontamination and an early condemnation of certain areas of the site [IRS 16].

4.3.2. *Public information or communication about the geological repository project*

Society has high expectations of both reliable information and active participation in decisions concerning its future and that of its descendants. A large number of failures in the implementation of large technical projects are linked to the failure to meet these expectations.

The NEA [AEN 06a; AEN 06b] has on several occasions engaged in reflection on this subject. Among the conclusions it has reached, it notes that in order to achieve a broad consensus, any decision-making process must respect three main principles. The first principle is that decision-making must be carried out through iterative processes, which allow sufficient flexibility to adapt to the evolution of the contextual situation, for example, by adopting a step-by-step approach that leaves enough time to develop a fair and acceptable discourse. The second principle is that collective learning should be facilitated, for example, by encouraging interactions between different stakeholders and experts [AEN 05]. Finally, public participation in the decision-making process must be encouraged, for example, by developing constructive and rich communication between individuals with different cultural backgrounds, beliefs, interests, values and views of the world.

Communication has a specific role to play in the development of deep geological repositories. Building trust with the stakeholders involved in this process, particularly within the local community, is essential for effective communication between authorities and the public. The NEA has compiled lessons from failures and successes in communicating technical information to non-technical audiences [NEA 17a].

4.3.3. *Measures to support a radioactive waste management project*

In a proposed radioactive waste management facility, it is important to demonstrate to the local population that the project will add functional, cultural and physical value to them (Table 4.2).

Functional aspects	Cultural aspects	Physical aspects
Multifunctionality or versatility	Originality	Integration
Adaptability	Aesthetic quality	Agentiment
Flexibility	Understanding	Accessibility
	Memorization	

Table 4.2. *Design features that contribute to maximizing the value added to a community by a radioactive waste management facility (source: [AEN 07a])*

The key to a lasting and positive relationship between a facility and its host community is to ensure that solutions are found together throughout the process. The best way to foster a sustainable relationship between a waste management facility and its host community is to add value to the project. This can be done through project design and new innovations. Several countries have taken this approach to welcoming communities to their sites in the short and long-terms [NEA 15c].

4.3.4. *Public participation in the geological repository project*

National policies for the long-term management of radioactive waste have been driven for decades by technical experts. The continuation of these technocratic policies has led in many countries to conflict with affected communities. Since the late 1990s, however, there has been a shift towards more participatory approaches. Despite these developments, technical aspects are still most often introduced into the public arena only after technical experts have defined the "problem" and decided on a "solution". Four countries, Belgium, Slovenia, Sweden and the United Kingdom, have made the participatory shift [BER 15].

The NEA updated a Short Guide to Stakeholder Engagement Techniques in 2015. This guide includes an annotated bibliography that assists practitioners and non-specialists alike by describing the steps and issues involved in involving stakeholders in decision-making and by facilitating access to useful online resources (manuals, toolkits and case studies) [NEA 15a].

A large-scale technical project can be managed according to two criteria, the orientation (internal or external) and the project structure (flexible or controlled). Depending on the choice of criteria, four relatively antagonistic models emerge. These models are human relations, internal process, open system and rationality of objectives models (Quinn *et al.* 1990 in [AEN 07b]). Each of these models has its advantages and disadvantages. The internal process model is more stable; in the case of the open model, resource acquisition and growth are possible. In the case of the objective rationality model, the advantages are productivity, efficiency and profit, while the human relations model allows for the development of human resources.

However, past experience clearly shows that large-scale technological projects are more likely to be approved when the various stakeholders have actively participated in their development and have developed an interest or responsibility in them. The trend in OECD/NEA countries is towards the development of forms of public participation that require the establishment or strengthening of a dialogue between all parties concerned [AEN 12a].

It is crucial to understand the dynamics of stakeholders and their impact on project management, especially for large complex projects such as nuclear waste repositories. To analyze changes in the negative or positive position (opposition or support) of stakeholders on a project, Aaltonen *et al.* [AAL 16] propose a new conceptual framework with a matrix of stakeholder positions according to their involvement in the project. They apply this concept to two examples: Onkado, which worked well, and Yucca Mountain, which met with a great deal of opposition.

4.3.5. *Information for future generations*

In order to avoid future human intrusion, the repositories containing the waste must be clearly identified in a way that is understandable to future society. The message delivered by the source (us in the 21st century) must be understood by the recipient (a population more than 10,000 years from now) (A). The challenge is great. There are three ways to classify signs (B) according to their dependence on context and the complexity of the information they can convey, symbolic (C), indexical (D) or iconic (E) (Figure 4.6).

The Nuclear Energy Agency launched the Preservation of Records, Knowledge and Memory (RK&M) Initiative across Generations in 2011. The main purpose of this initiative was to foster international thinking and progress in this direction and to respond to the growing demands from waste management specialists and other interested parties for viable and common strategies. Phase I concluded on September 15 and 17, 2014, with the organization of an international conference and debate on "Building Memory," held in Verdun, France. The second phase ended in 2017 [NEA 15b].

To achieve this goal, the initiative focused on five key issues:

– For what reasons and purposes do we need and want to preserve archives, knowledge and memory about radioactive waste across generations?

– What type of information should be kept?

– On what time scales?

– By whom and for whom?

– What can be done now and in the future to ensure maximum continuity and accessibility to RK&M? [NEA 19]

In the past, RK&M's preservation efforts have focused on preventing inadvertent human intrusion through hazard-based messaging and methods and promoting aversion. While deterring potential intruders remains a valid goal, it was found that this should instead be achieved by supporting an informed and alert attitude towards required levels of safety, security and societal compliance. Achieving these goals can only be done in a manner that combines the preservation of records, knowledge and memory. Indeed, it is not only a matter of transmitting a message, but of keeping it interpretable, meaningful, credible and usable over time. The results and projects of this initiative are numerous, and a list of 35 mechanisms has been compiled, grouped into nine general approaches [NEA 19].

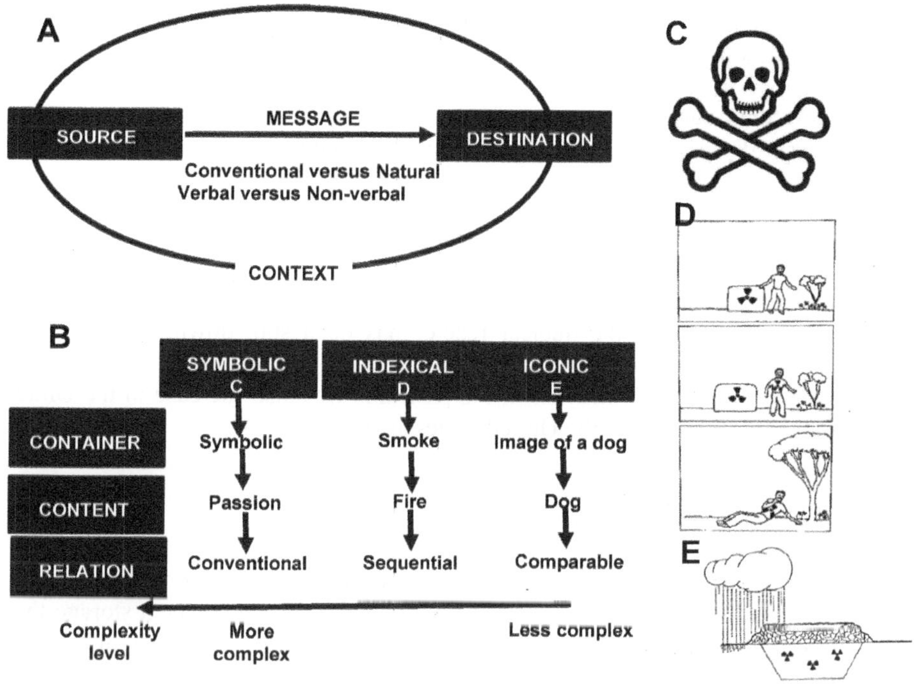

Figure 4.6. *Type of message in relation to the context (A) and classification of signs and relationship between the container (signifier) and the content (signified) of the message (B). Examples of messages: (C) the skull and crossbones symbol, Jolly Roger; (D) the indexical message showing the effect of radiation on humans; (E) the iconic sign showing the radioactive waste site (according to Home, in [IAE 20]). For a color version of this figure, see www.iste.co.uk/amiard/radioactive.zip*

Faced with the challenge of the large volumes of RK&M materials generated by national disposal programs, it became apparent that a suitable response was the concept of a Key Information File (KIF). The KIF is designed to be a single, short document, produced in a standard format, with the goal of enabling society to understand the nature and intent of a repository, and thus reduce the risk of unnecessary human intrusion. Three draft key information files being developed to support the preservation of RK&M in France, Sweden and the United States are provided as examples [NEA 19].

4.4. Alternative solutions

The volumes of high-level radioactive waste, such as spent fuel, are increasing very rapidly in all nuclear countries. Temporary storage facilities are approaching saturation in many countries. There is an urgent need to find a pragmatic solution. As final storage projects are long and delicate to initialize, it is becoming beneficial to opt for other solutions, even though they are only temporary.

The solutions currently envisaged are medium-term temporary storage either under water or dry, long-term storage (100 years) or deep continental drilling.

4.4.1. *Underwater temporary storage*

In general, states that opt for a closed nuclear fuel cycle temporarily store their spent fuel in pools, after the nuclear reactor has been shut down, and then in a centralized site for reprocessing. This is the case, for example, in France at La Hague and in Russia at Mayak. In the past, this was also the case in the United Kingdom at Sellafield and in the United States at Hanford.

4.4.2. *An interim solution: dry storage*

After being used in nuclear reactors, irradiated fuel is stored for a few years in cooling pools adjacent to the reactors. The question of more permanent storage then arises, when their radioactivity and the heat that they give off have sufficiently diminished. In the majority of countries with nuclear power plants (the United States, Germany, Sweden, Japan, South Korea, etc.), the solution of dry storage has been chosen [LAP 18].

4.4.3. *A waiting stage: long-term storage*

Long-term storage will necessarily have degradation and corrosion effects on the spent nuclear fuel and high-level waste that will interfere with subsequent geological disposal. These potential long-term degradations of these systems will have to be taken into account when applying for a storage license renewal. Indeed, long-term degradation of storage and transport systems could result in non-compliance with storage and/or transport requirements. Therefore, long-term aging management plans must be developed [KES 17].

4.4.4. *The American perspective of deep drilling*

The first American studies date from the 1970s [SCH 74e, BRE 78]. Waste isolation in Very Deep Hole (VDH) boreholes is an alternative with wells of about 50 cm (20 inches) diameter that are drilled by modified rotary rigs and at depths of about 6,000 m (20,000 feet) or less [ONW 83].

The British company NIREX has carried out a review of the development of the concept of deep borehole disposal of high-level radioactive waste [NIR 04]. It notes that most of the work in this field, with respect to the disposal of HLW and spent fuel, has been carried out either for the US Department of Energy (USDOE) or by SKB. In addition, this disposal concept was also considered, mainly in the United States, for the disposal of excess weapons-grade plutonium. However, there were great uncertainties in understanding the geology, hydrogeology and hydrogeochemistry of the rocks at great depth. It was not until the SKB program began in the 1980s that serious consideration was given to what conditions might be at depth in both crystalline basement and deep sedimentary rocks and what the major benefits would be of using this concept for long-lived waste storage. SKB had considered several options, such as KBS-3, WP-Cave, medium to long boreholes and very long boreholes.

More recently, there has been renewed interest in the use of the deep drilling concept, with alternatives to the normal definition of the concept being suggested, including, for example, melting or partial melting of the host rock by waste or reuse of former hydrocarbon wells [NIR 04].

The Americans still have the prospect of disposing of nuclear waste in deep wells. Deep drilling is almost technically feasible today. According to Brady *et al.* [BRA 17], characterization, drilling and placement of radioactive waste at a site could probably be accomplished within 5 years, or sooner, and could be less costly than disposal in a mined repository while being just as safe.

4.5. Management of high-level radioactive waste by the various States

According to the International Atomic Energy Agency (IAEA), at the beginning of 2003, there were some 171,000 tons of spent nuclear fuel from nuclear power plants in interim storage worldwide. Of this total, about 36,000 tons were in Western Europe and about 28,000 tons in Eastern Europe. The total amount of spent nuclear fuel planned for the world in 2010 was expected to be about 340,000 tons [MEA 07]. In the European Union, 21 states manage about 59,000 tML of spent fuel, and each year, about 3,200 tML of additional spent fuel are generated (Martin-Ramos *et al.*, in [IAE 20]).

The three most advanced countries in the world for deep geological disposal are Finland, Sweden and France. In these countries, a site has already been selected and the associated construction applications have been accepted (Finland), submitted and reviewed pending a final decision (Sweden) or about to be submitted (France) [CNE 19a].

Sanders and Sanders [SAN 16; SAN 17] have summarized the policy of the majority of states. With respect to high-level and long-lived radioactive waste, national policies are very diverse. Thus, the types of containers and backfill vary from one country to another, whether for spent fuel (Table 4.3) or HLW (Table 4.4). Similarly, the expected functions of the arrays and containers are not identical across states (Table 4.5).

Country	Container	Backfill	Other
Canada	Carbon steel, copper	Bentonite, sand, clay, rocks	Tunnel and joints
South Korea	Steel, copper	Bentonite, sand	-
Spain	Carbon steel	Bentonite	Concrete, bentonite
United States (YMP)	Stainless steel, Nickel-based alloy	-	Concrete, joints
Finland	Copper, iron	Bentonite, rocks	Bentonite, concrete
France	Stainless steel	Concrete pavement	Bentonite seals
Sweden	Copper, iron	Bentonite	Tunnel, backfill

Table 4.3. *Spent nuclear fuel (SNF) waste conditioning in various countries (source: [OJO 14])*

Country	Matrix	Container	Backfill	Other
Belgium	Glass	Stainless steel	Clay, Bentonite, quartz, graphite	Tubes in tunnels
China	Glass	Stainless steel	Bentonite	Bentonite seals
Czech Republic	Glass			Clay joints
France	Glass	Stainless steel	Bentonite	Bentonite, concrete joints
Japan	Glass	Carbon steel	Bentonite, sand	Tunnel with sealing and grouting
Russia	Glass	Stainless steel	Bentonite, concrete	Bentonite, concrete
Switzerland	Glass	Steel	Bentonite	
United States (YMP)	Glass			

Table 4.4. *Conditioning of high-level waste (HLW) in various countries (source: [OJO 14])*

Country	Function of the matrix	Function of the container
Belgium	10,000 years of leaching resistance	Easy handling
Canada	10,000 years of retention of radionuclides	Contain 100,000 years
South Korea	Leaching resistance	Contain 1,000 years
Spain	Low release rate	Contain 1,000 years
United States (YMP)	Reduced release rate	Corrosion resistance >10,000 years
Finland	Low release rate	Contain 100,000 years
France	100,000 years of resistance	Easy handling
Japan	Low salting out, resistance >10,000 years	Contain 1,000 years
Czech Republic	Retention of radionuclides 10,000 years	Contain 500–1,000 years
United Kingdom	Limit release 300–500 years	Integrity 300-500 years
Sweden (KBS-3)	Low release rate	Contain 1,000 years
Switzerland	Low release rate 150,000 years	Initial containment

Table 4.5. *Matrix and container functions in various deep geological repository projects (source: [OJO 14])*

4.5.1. *States advocating a closed nuclear fuel cycle*

All nuclear weapon states have also developed reprocessing of spent nuclear fuel to extract plutonium. Only a few countries maintain civilian reprocessing plants with the hope of being able to close the fuel cycle in the future. These are China, France, India and Russia [IAE 20]. Japan also had this desire, but these reprocessing plants are all shut down with little hope of restarting.

This closed-cycle option has a major impact on the management of radioactive waste. Indeed, this means that all radioactive materials (new fuel, spent fuel, extracted uranium and plutonium) are not considered as waste and are therefore not counted as such.

4.5.1.1. *China*

China is committed to a closed-loop nuclear fuel cycle strategy with reprocessing. China has expressed interest in building a deep geological repository for the storage of high-level waste from its nuclear facilities. However, the overland transportation of spent fuel via state-owned highways, more than 4,000 km from nuclear power plants in the eastern provinces to designated interim storage facilities and then to the final storage site, appears to be both the bottleneck and the Achilles heel of the nuclear waste strategy.

The quantity of spent fuel will be about 12,300 tons in 2030. The projected timetable for the geological repository is to choose the site and configuration before 2020, from 2020 to 2040, the construction of the underground laboratory and *in situ* tests, and from 2041 to 2050, the construction of the repository itself [STE 18a].

4.5.1.2. *France*

In France, the filling rate of fuel building pools remains high, even though it has improved in recent years according to EDF [EDF 16]. This is not the opinion of the ASN [ASN 18b], which considers that the storage pools for nuclear power plant fuel (BK pools) and the pools at the La Hague site have little margin before saturation. It is linked in particular to the capacity of the La Hague and MELOX plants to process spent fuel and to manufacture MOX (Mixed Oxide Fuel) assemblies, using plutonium from reprocessing. EDF is also studying the construction of a centralized pool dedicated to spent fuel (ERU and MOX) in order to increase storage capacity for a hundred years. This pool, consisting of two independent 5,000 t pools, should meet a high level of safety requirements to take into account the lessons learned from Fukushima. This pool could be put into service by 2028–2030. It would provide operating and safety margins by reducing the number of pools in the nuclear power plants and in the La Hague plant.

Centralized dry storage of spent fuel is possible because the CEA has been using a dry storage casemate for spent fuel (CASCAD) at Cadarache since 1990. This storage facility, with 319 wells, holds fuel from the heavy water reactor EL4 at Brennilis [IRS 18a].

The possibilities of dry storage of MOX fuel also exist. Indeed, the characteristics of the MOX fuels used by EDF have evolved over time. In particular, their plutonium content has successively been 5.30%, 7.08% and 8.65% (current content). Based on current concepts, for all spent MOX fuels with a plutonium content equal to 5.30% and for those with a content of 7.08%, about 2,500 assemblies could be envisaged as of now. On the other hand, for the first spent MOX fuels with a plutonium content of 8.65%, used since 2007, it would be necessary to wait until around 2040.

About 1,150 ERU fuels were loaded into reactors by EDF between 1994 and 2013. All the spent ERU fuels currently in storage have a thermal power of less than 2 kW and are therefore compatible with current dry storage concepts [IRS 19a].

4.5.1.3. *India*

India is a state strongly committed to waste recycling and works for many countries around the world. Among the large mass of waste are solid radioactive waste [BAN 19] and hospital waste, some of which contains radioactivity [TIW 13]. Radioactive waste has been discovered in New Delhi. However, the government is reluctant to impose standards on this informal sector, which employs tens of thousands of people [BOU 10].

From a nuclear point of view, India has been a nuclear country for many years, both on a military and civilian level, with a complete nuclear fuel cycle. Its future program is important and it is engaged in the construction of new power reactors (2 x 700 MWe) [SEE 17].

The Department of Energy (DOE) is responsible for radioactive waste from nuclear weapons production and certain research activities. The Nuclear Regulatory Commission (NRC) regulates the management, storage and disposal of commercial radioactive waste from power generation and other non-military uses of nuclear materials. The NRC has entered into agreements with 32 states, called Agreement States, to allow those states to regulate the management, storage and disposal of certain nuclear waste [THA 15].

In India, the Atomic Energy Regulatory Board (AERB) published a dozen guides on radioactive waste management in 2006 and 2007. The AERB develops recommendations for the surface disposal of solid radioactive waste. It also provides specific recommendations for the use of radioactive sources, sealed or not, in medicine, research and industry and for mining waste [AER 07a].

LLRW and ILRW is stored either in earthen trenches, in trenches reinforced with concrete, or in cavities lined with porcelain tiles. The main storage centers and management facilities for Indian radioactive waste are on the coast, Trombay (Maharastra, 1956), Tarapur (Maharastra, 1969) and Kalpakkam (Tamilnadu, 1984), and inland, Kota (Rajasthan, 1972), Narora (Uttar Pradesh, 1989), Kakrapar (Gujatar, 1993) and Karga (Karnataka, 2000) [AER 06].

As for all states, the primary objective of radioactive waste management is the protection of human health, the environment and future generations. The general philosophy of safe radioactive waste management is based on the concepts of (i) latency and physical decay, (ii) dilution and dispersion and (iii) concentration and containment.

India's classification of radioactive waste includes three classes: LLW, ILW and HLW. India has achieved self-sufficiency in the management of all types of radioactive waste resulting from the operation of all nuclear fuel cycle facilities. India manages all waste from uranium mining, fuel fabrication to reactor operation and subsequent reprocessing of spent fuel. Since spent fuel is reprocessed to recover and reuse the uranium and plutonium produced, the fuel cycle is described as "closed" [WAT 13].

As a national policy, near-surface disposal facilities (NSDFs) are located at each nuclear facility site.

For HLW, India is seeking to reduce its long-term radiotoxicity and, to do this, is developing a strong research program on the separation and transmutation of long-lived minor actinides (Np, Am, Cm) and fission products (^{129}I, ^{99T}c, etc). For this theoretical transmutation, India would use integral fast reactors (IFRs) or accelerator-driven sub-critical systems (ADSS), leading to the elimination or reduction of radioactive inventories [WAT 13].

4.5.1.4. *Russia*

Information on the management of Russian radioactive waste is not very accessible. The dismantling of Soviet submarines, carried out largely through international cooperation, has produced large quantities of radioactive waste [AMI 21]. Thus, the quantities of spent fuel are of civilian origin 8,740 tons ($2.3.10^{20}$ Bq), of military origin 30 tons ($5.6.10^{19}$ Bq) and of Ministry of Transport origin 10 tons ($1.7.10^{18}$ Bq) [JAI 18]. In addition, the Russians have to manage the waste from the accidents that occurred at Mayak.

The Russian reprocessing plant RT-1 has been operating since 1977 and to date more than 6,000 tons of spent fuel have been reprocessed. In addition, Russia operates two industrial-scale fast reactors (BN-600 and BN-800) giving it some assurance for the closed option of its nuclear cycle (Khaperskaya *et al.*, in [IAE 20]).

The development of the necessary infrastructure, the increased reprocessing of spent nuclear fuel and the establishment of a closed fuel cycle must provide answers to the pressing problem of radioactive waste disposal. This is all the more crucial as the Russians wish to commercialize the management of foreign radioactive waste.

The project for Russia's first underground repository for HLRW and ILRW will be constructed in the Archean crystalline granite and gneiss at Yeniseisky (Krasnoyarsk region, Siberia). The HLW will be disposed of as aluminophosphate glass, and the ILW (with long-lived radionuclides) will be cemented. Preliminary research on all aspects of the repository design (stability of waste forms, waste packages and bentonite buffer, evaluation of the geological barrier and simulation of radionuclide transport by groundwater) will be carried out in an on-site underground research laboratory [LAV 16].

In June 2001, despite public opposition, the Duma approved a bill sponsored by President Putin and Russia's powerful atomic energy ministry, Minatom, that would allow Russia to absorb up to 20,000 tons of spent fuel from countries such as Taiwan, South Korea, India and Japan over the next 10 years, generating up to $20 billion in much-needed cash flow [ONE 02].

4.5.2. *States that have reprocessed spent fuel in the past*

Several countries have in the past reprocessed their spent nuclear fuel either for military or civilian purposes. This was the case for Germany, South Africa, Belgium, Brazil, the United States and the United Kingdom. They have all abandoned this technique and thus the closed nuclear fuel cycle. The closure of Sellafield in the United Kingdom was very recent (end of 2018). Some states continue to reprocess nuclear fuel but only for military purposes, such as Pakistan, Israel or North Korea, with some doubts for the last state.

Two states have a more unique history. Japan has in the past practiced reprocessing and had plans for new plants. But all its facilities have been shut down for a long time, and experts are wondering whether this is a temporary or permanent withdrawal. The second state is in the opposite situation. South Korea has never carried out reprocessing, but wants to use it to relieve the pressure on its spent fuel storage facilities.

4.5.2.1. *South Africa*

In South Africa, a research reactor was commissioned in 1965 and has been producing nuclear electricity since 1984. This country had a secret nuclear weapons program between 1978 and 1990. There are about 3,000 irradiated fuel assemblies in South Africa, some from nuclear reactors (Koeberg-1 and 2, Castor X/éSF) and some from the former military program (Pelindaba) [FIG 18].

4.5.2.2. *Germany*

In Germany, the quantity of spent fuel at the end of 2001 was 14,645 tons; with dismantling, the total will be 17,416 tons. Reprocessing will concern 6,795 tons, and 10,567 tons will be directly stored [HOC 15a; HOC 15b].

Germany has centralized storage facilities at Gorleben and Ahaus where spent nuclear fuel from PWRs and BWRs is isolated in dual-use containers (transport and storage) [IRS 18a].

For geological disposal of radioactive waste, Germany has conducted 40 years of repository research and more than 500 research projects. This has provided the scientific basis for a safe repository for high-level waste in rock salt as host rock. The chosen site is the Gorleben salt dome. The creep capacity (irreversible deformation) of the highly ductile rock salt will lead to the rapid closure of the mine openings and fill the storage areas, leading to the containment of the deposited waste. The very low fluid content of the salt will limit chemical processes such as corrosion of the waste packages [VON 17].

4.5.2.3. *Belgium*

In Belgium, the geological disposal project for intermediate- and high-level waste is the subject of a tricky process. In 1974, one year before the first Belgian reactors were put into operation, the Belgian Nuclear Energy Study Centre (*Centre d'Étude de l'Énergie Nucléaire belge,* SCK-CEN) decided to begin design studies for a geological repository for B and C waste (AA-LL and HLW) and to select a site.

Geological studies and several boreholes have shown that below the SCK-CEN site in Mol there is a layer of so-called "Boom clay". This layer was slightly indurated, plastic, 100 m thick and about 250 m below the surface. It would therefore make it possible to consider the implementation of an underground laboratory on the site in order to study the properties of the clay as a host rock for a possible storage site. Construction of the laboratory began in 1980. Since 1995, the underground laboratory has been operated by EURIDICE, an economic interest grouping between SCK-CEN and ONDRAF.

Between 2011 and 2018, ONDRAF submitted several "national policy" proposals to its supervisory body, all of which were based on a geological storage solution on national territory. However, no policy concerning geological disposal has been clearly decided in Belgium up to now [CNE 19a].

The implementation of this project is a cautious and progressive process, punctuated by the submission of key documents to the government. Ypresian clays are considered an alternative to Boom clay [SMI 09].

According to Schroder *et al.* [SCH 15], the Belgian case of Category B and C waste governance presents an ambiguous combination of highly specialized advanced and focused research with rather impassive, fragmented and delayed policy making. For more than 30 years, radioactive waste management has been in a research stage. The focus has been on scientific progress in geological disposal in the Boom clay layer, aided by the experiences gained from the underground research laboratory in the municipality of Mol. The process is currently frozen.

Dual-use drums (transport and storage) have been the subject of numerous concepts since the 1990s (CASTOR and CONSTOR from GNS, TN 24 from TN International, HI-STAR 100 from HOLTEC International and several options from NAC-STC in the United States, among others). In Switzerland and Belgium, spent fuel is stored in metal containers of the TN 24 family from Orano TN, such as the TN 24 DH at Doel in Belgium [IRS 18a]. At Tihange, in Belgium, irradiated fuel is stored in a pool. All the spent fuel stored at the end of 2012 represented 10,394 assemblies and 4,711 tML [ZER 17]. Since reprocessing took place at La Hague in France, vitrified waste (returned to Belgium) and compacted metallic waste are still partly stored there (147 m^3) [ZER 17].

4.5.2.4. *South Korea*

Since the temporary storage facilities for spent fuel at nuclear power plants will reach their full capacity in the near future, the disposal of spent nuclear waste has become a matter of serious concern in South Korea.

The amount of spent fuel in storage is 13,423 metric tons, which is 71% of the total pool storage capacity. South Korea has studied a large number of possible sites for its high-level nuclear waste storage. Many of its projects have been abandoned for various reasons, often due to hostility from local populations. In 2005, the Gyeongju site was selected by the government as a site for nuclear waste disposal [LEE 18].

Storage at Wolsong is in the form of dry storage in Canadian-designed vertically loaded casemates.

South Korea, in collaboration with its military ally the United States, plans to reprocess its spent nuclear fuel using a new experimental pyro-processing technique.

4.5.2.5. *United States*

In the United States, high-level nuclear waste includes large volumes (tens of millions of cubic meters), high total activities (billions of curies) and very diverse and complex compositions [EWI 95].

US policy towards nuclear waste is detailed in a Department of Energy document [DOE 17]. Commercial spent nuclear fuel is temporarily stored in the vicinity of 75 reactors in 33 states in pools that provide cooling and radiation protection. The primary risk to the pools is the loss of cooling and shielding water.

The US pools have reached their capacity limits, and utilities have implemented dry storage.

Spent nuclear fuel (SNF) and HLRW managed by the DOE are being temporarily stored at five sites in five states. For HLW a total of about 20,000 containers are planned, and for spent fuel, a total of 2,458 tons of heavy metal (2,149 of which are from DOD) or about 3,500 containers (Table 4.6).

Voegele [VOE 17] published an inventory with 221,000 commercial spent fuel assemblies, 22,475 containers of HLW or a total of about 70,000 tons of heavy metal.

Sites	**High-level waste (number of containers)**	**Spent nuclear fuel (SNF) (in tons of metal)**
Hanford	9,700	2,130 (2 102)
Idaho	3,500–5,000	280 (36)
West Valley	275	
Savannah River	2,900–6,300	30 (10)
Fort St-Vrain		15
Other sites		2 (<1)
Total	3,125 (in 2010) 19,865–21,365	2,458 (2,149)

Table 4.6. *High-level waste that was in storage in the United States in 2010 and the projection into the future (source: [OSB 18]). Military waste in parentheses*

In 2018, the projected tonnages of spent nuclear fuel and HLRW in the United States were approximately 80,150 tons of heavy-metal spent fuel stored in the United States for 25,400 tons of heavy-metal dry storage at reactor sites in approximately 2,080 drum systems, with the balance in pools, primarily near reactors. Dry storage systems for spent nuclear fuel are in dual-purpose canisters (DPC).

Approximately 2,200 tons of heavy-metal spent fuel will be generated nationwide each year in the United States and approximately 160 new dry storage containers will be required. By 2048, the waste volumes will be 183,896 (85%), 25,555 (12%) and 7,165 m^3 (3%) for commercial spent nuclear fuel, HLW and military spent fuel, respectively [OSB 18].

Spent fuel is stored dry in HOLTEC's vertical-loading casemate system at the San Onofre Nuclear Generating Station, Humbolt Bay Power Plant and Ameren's Nuclear Power Plant sites.

The United States has two deep geological repositories, one in operation for military radioactive waste, the Waste Isolation Pilot Plan (WIPP), the other is in the pipeline, at Yucca Mountain.

WIPP began operation in March 1999 in southeastern New Mexico. It is a storage center for transuranic military waste, located at a depth of 635 m, in a salt layer. This pilot plant is operated by the US Department of Energy (DOE). The USEPA specifies that the DOE must demonstrate on a sound basis that the WIPP disposal system will be effective in containing the alpha emission of long-lived radionuclides in its enclosures for a minimum of 10,000 years after closure of the storage [LAR 00]. The EPA believes that the long monitoring has resulted in good characterization of key parameters.

For the Yucca Mountain underground storage project, although the review is complete and the safety assessment report has been issued, the decision of the Atomic Safety and Licensing Committee remains pending. The Nuclear Regulatory Commission has issued a draft supplemental environmental impact statement to address questions raised by staff at the time of the safety analysis report. A final environmental impact statement was issued in early 2016.

The Department of Energy's program for HLRW disposal has been hampered by overwhelming political opposition fueled by public perceptions of great risk. Analysis of these perceptions shows that they are deeply rooted. Consequently, it is unlikely that the choice of site and development of a national repository for radioactive waste disposal will be feasible in the foreseeable future [SLO 91]. Similarly, according to Forrest [FOR 15], the US Yucca Mountain project has been halted.

4.5.2.6. *Japan*

The management of radioactive waste in Japan is not clearly decided. It is still discussed separately from nuclear policy in Japan. Electricity companies use nuclear power without serious concern for nuclear waste. Even after the Fukushima nuclear accident, they tried to restart their reactors as soon as possible (in 2020, only 9 reactors out of 54 had restarted). Since the Fukushima accident, Japanese public opinion has been overwhelmingly hostile to the continuation of nuclear power generation.

Moreover, Japan still seems to have the will to reprocess its spent nuclear fuel. However, all of its reprocessing facilities are in extended shutdown or construction is suspended. Currently, part of its radioactive waste is stored abroad, including France. In addition, Japan has decided to abandon the Monju fast reactor, which has been plagued by problems for years, but it is not abandoning its quest for a fast demonstration reactor.

In 2014, Japan stored 14,870 tons of spent fuel with a total capacity of 20,740 tons. The cost of waste management was estimated in 2011 according to three options; all reprocessed about €0.0149 per kWh, half reprocessed €0.0105 per kWh and no reprocessing €0.0075 to €0.0079 per kWh [OKA 18]. The situation of fuel assemblies (irradiated and new) is of concern at the Fukushima site following the 2011 accident. In 2019, 4,999 fuel assemblies remained in the pools of reactors 1, 2, 3, 5 and 6 (the pool of reactor 4, which was badly damaged, has been evacuated), 6,105 fuel assemblies in a common pool and 2033 fuel assemblies temporarily stored dry (Shui *et al.*, in [IAE 20]).

Japan has a project for a centralized dry storage facility with packages at Mutsu pending reprocessing of spent fuel [IRS 18a].

In May 2016, the Nuclear Energy Agency conducted an independent peer review of the siting process for geological disposal of HLRW in Japan. It provided detailed remarks for the use of clear terminology in the screening process for siting [NEA 16c]. Japan is disadvantaged by its geological situation with many areas of high seismic risk, rugged terrain and population density on the coastline. Moreover, the island of Hokkaido is a nuclear-free zone.

Regarding the Japanese reprocessing plant project at Tokai with a processing capacity of 800 t and an annual production of 8 t of plutonium, the JNFL (Japan Nuclear Fuel Limited) announced on September 30, 2017, a new delay, the 24th postponement. The estimated cost of this plant is 25.7 billion dollars, four times the initial forecast. Tepco Group (operator of the destroyed Fukushima plant) and Japco are building a dry storage facility for 3,000 tons of spent fuel in Mutsu, a town near Rokkasho. A second pool had to be built due to delays in the construction of the plant. At the end of September 2015, there were 14,730 tons of spent fuel in storage at Japan's 17 nuclear power plants and 4,820 tons inside the reactors [LAP 17].

Japan is facing an acute radioactive waste management problem following the multiple nuclear accidents at Fukushima in 2011. There are four types of waste management selected for the Fukushima waste. These are trenching, pits, intermediate-depth disposal and deep geological burial.

Radioactive waste is divided into two categories, 1 and 2. Category 1 includes waste with the highest concentration (higher than that of category 2) and its storage is obligatory in deep geological disposal. No technical requirements concerning the waste packages and the deep geological disposal have been settled.

Category 2 is divided into three subcategories: trench, in-pit and intermediate depth. Artificially covered trench disposal is a method of burying radioactive waste at or below ground level (below the surface to a depth of 50 m) where the concentrations of radioactivity of three radionuclides do not exceed a certain limit (Table 4.7). Buried pit disposal is a method of burying radioactive waste below ground within 50 m of the surface where concentrations of seven radionuclides do not exceed certain limits. Intermediate-depth disposal is at a depth greater than 50 m, and this subcategory is defined by the concentration limits of four radionuclides (Table 4.7).

Management (Depth) Radionuclide	**Trench (0–50 m)**	**Pit (0–50 m)**	**Intermediate depth (> 50 m)**
Cobalt-60	10 GBq/t	1 PBq/t	
Strontium-90	10 MBq/t	10 TBq/t	
Cesium-137	100 MBq/t	100 TBq/t	
Carbon-14		100 GBq/t	10 PBq/t
Nickel-63		10 TBq/t	
Technetium-99		1 GBq/t	100 TBq/t
Alpha emitters		10 GBq/t	100 GBq/t
Iodine-129			1 TBq/t
Chrome-36			10 TBq/t

Table 4.7. *Mass activity limits per ton of waste for various radionuclides characterizing three subcategories of Category 2 waste with three management types as a function of burial depth (source: [NEA 16a])*

4.5.2.7. *United Kingdom*

In the United Kingdom, high-level waste includes intermediate-level waste, which is currently stored on nuclear sites, particularly at Sellafield, until a disposal solution is found. LLW is disposed of in a number of ways, including burial in the national facility at Drigg (County Cumbria), 6 km southeast of Sellafield.

Since the 1940s, the United Kingdom has accumulated a substantial amount of radioactive waste from civilian and military nuclear activities. This amount is expected to increase over the next century as existing nuclear facilities are decommissioned [CNE 19a]. The total volume of conditioned waste to be stored is currently estimated at about 740,000 m^3. This conditioned waste is divided into vitrified HLW (10,000 m^3), ILW (460,000 m^3), LL-LLW (11,000 m^3), waste contaminated by plutonium (620 m^3), spent fuel (68,000 m^3) and depleted or low-enriched uranium (190,000 m^3). All of this waste is currently stored on about 30 sites, but mainly on the surface at Sellafield (Cumbria) [CNE 19a].

Type of waste	Volume (%)	Activity (millions of TBq)	Activity (%)
HLW	< 0.3	39	50
ILW	73.9	2.4	3
Separated plutonium	0.7	4	5
Uranium	15.7	3	<0.01
Spent fuel	1.7	3.3	42

Table 4.8. *UK waste inventory excluding LLW (source: [MAC 15])*

For long-lived ILRW and HLRW, there is currently no solution in the United Kingdom. Several attempts to locate geological disposal facilities for high-level waste have already taken place. From the 1980s to the 1990s, initial efforts were aimed at selecting a site for intermediate-level waste through a top-down approach, entirely carried out behind closed doors. This led to an application for an underground rock characterization facility near Sellafield. The application was rejected on appeal in 1997 [CNE 19a].

In 2001, the government began a new public consultation process on the disposal of high-level waste. To gain public support, it launched a broad public debate, overseen by an independent body, the Committee on Radioactive Waste Management (CoRWM). Public participation was significant, and the review of the many disposal options took nearly five years. This long delay was widely criticized. The committee's final report, published in July 2006, led to two major recommendations: long-term disposal of high-level waste should be carried out in a deep disposal site, a solution called "geological disposal", and reliable interim storage should be provided because the development of the final site could take several decades [MEA 07].

A subsidiary of the NDA (Nuclear Decommissioning Authority), the RWM (Radioactive Waste Management), is specifically responsible for managing the UK's civilian radioactive waste. The RWM uses a consensus-based approach and has initiated a second attempt to identify suitable sites that was conducted from 2008 to

2013. The methodology followed is described in a White Paper on the Safe and Sustainable Management of Radioactive Waste [DEF 08] and shown in Figure 4.7. The process gives the local communities and the RWM a permanent right to withdraw from the project. The process was closed in January 2013 after one of the three local authorities involved in Cumbria decided not to proceed to the next decision-making stage [CNE 19a].

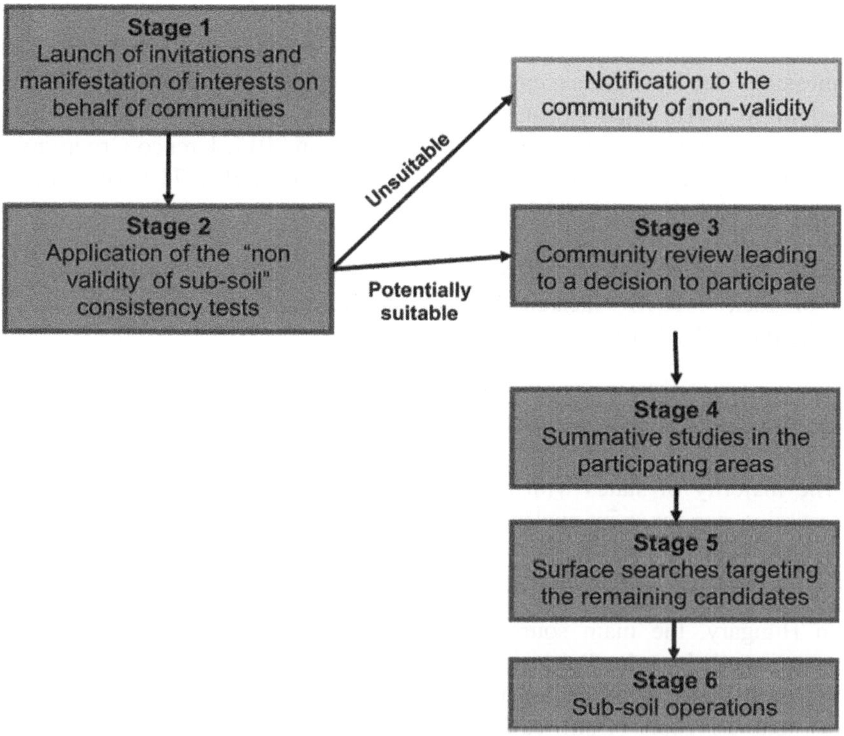

Figure 4.7. *Stages in the site selection process in the United Kingdom (source: [DEF 08, AEN 10a]). For a color version of this figure, see www.iste.co.uk/amiard/radioactive.zip*

While waiting for a final solution, temporary storage is being organized. For example, the solution that EDF Energy is putting into practice at the Sizewell B nuclear power plant that it operates is as follows. Electricity production at Sizewell B began in 1995, and the spent fuel has been stored until now in a pool, which was not designed to store spent fuel for the entire operating life of the plant. The choice of a medium-term solution (up to 100 years) was made in favor of dry storage, pending the availability of final storage. The project started in 2009, and its construction in 2013. The fuel was placed in stainless steel containers, welded and placed in drums containing 120 tons of shielding and concrete. Construction of the building and testing of the unirradiated fuel was completed in 2016. The first drums of irradiated fuel were transferred there in the first half of 2017.

The general British strategy is to use the fuel in the AGR reactors to the maximum, prior to their scheduled shutdown from 2023. This will make it possible to smooth out evacuations to the Sellafield storage center [EDF 16].

Uranium enrichment company Urenco said in September 2017 that it plans to commission the Tails Management Facility (TMF) at Capenhurst in the United Kingdom in 2018, after construction delays. The facility, to convert depleted uranium hexafluoride tailings, entered into service in 2017. Urenco Group invested €151.2 million ($179.5 million) in the first half of 2017, 71% of which was associated with tailings [WNN 17].

Forty-five years after its initial design and after reprocessing more than 9,300 tons of fuel, THORP sheared its last fuel assembly on November 9, 2018. (Hallington, in [IAE 20]).

4.5.3. *States with an open nuclear fuel cycle*

The majority of states with a civilian nuclear industry have not developed a reprocessing step for spent nuclear fuel and therefore have an open fuel cycle.

4.5.3.1. *Hungary, Lithuania, Slovenia and Ukraine*

In Hungary, the main source of radioactive waste is the four WWER-440 reactors installed at the Paks site. They began producing commercial electricity between 1982 and 1987. The quantities of spent fuel are 1,095 tons of commercial origin and 267 tons from education and research. Currently, the spent fuel assemblies have been placed in the Spent Fuel Interim Storage Facility (SFISF) since 1997, which is located in Paks. It is a Modular Vault Dry Storage (MVDS) facility (Fisker and Geher, in [IAE 20]).

The reference scenario calls for deep geological storage, but the possibility of shipping the spent fuel to Russia for reprocessing remains open. Spent fuel is currently stored in an interim storage facility next to the Paks nuclear power plant. For deep storage, the most favored site is the Boda Siltstone formation in the western Mecsek Mountains in southern Hungary [KOR 18].

In Lithuania, there is only one nuclear power plant, the Ignalina nuclear power plant. Its two units are permanently shut down and being decommissioned. From 1983 to 2010, about 21,500 spent nuclear fuel assemblies were accumulated. Since

2003, two near-surface storage repositories (20 m deep) are being installed near the Ignalina NPP. These are a landfill for 60,000 m^3 of very low-level radioactive waste and a near-surface repository with engineered barriers for 100,000 m^3 of low- and intermediate-level radioactive waste [POS 18].

When the Ignalina plant was transferred to Lithuanian control in 1991, 90% of the storage ponds were already in use. A solution had to be found quickly in order not to close the plant prematurely. Lithuania therefore decided to build an interim dry storage site near the plant, designed to last more than 50 years, dry storage in the form of dual-use packaging (CASTOR process). This process involves sealing the spent fuel inside concrete and metal (black steel) containers. The site was initially designed to store 72 containers. The dry surface storage site was opened in 1999. It should be noted that studies have shown that the geological situation of Lithuania does not allow for the construction of a repository in deep rock to store radioactive waste [MEA 07].

In Slovenia, there was no mention of nuclear waste in the plans for the first Yugoslav nuclear power plant in Krško drawn up in the 1970s. Nor was there any mention of it in the purchase contract with Westinghouse. The first spent fuel management strategy was adopted in 1996. It was based on an agreement between the governments of Slovenia and Croatia. However, cooperation proved difficult, as each side wanted to gain the maximum benefit from the nuclear power plant and bear the least possible cost for the spent fuel. Two separate final repositories for nuclear waste were likely [ŠEŠ 18]. The program to develop geological repositories was suspended in 2004 [MIH 18].

Today in Ukraine, radioactive waste from the Chernobyl NPP site and the surrounding exclusion zone constitutes more than 98% of the total solid radioactive waste. After Ukraine's independence from the Soviet Union, its institutional system for managing nuclear waste problems has constantly changed and has not reached the stage of clearly defined responsibilities and distribution of roles among different institutions. Due to the ongoing military conflict with Russia, Ukraine has lost control of its research reactor in Sevastopol and its nuclear waste collection center in Donetsk [PAS 18].

4.5.3.2. *Canada*

In Canada, in 2015, there were 1,102,470 spent fuel bundles in dry storage and 1,496,018 bundles in wet storage [VES 18].

Spent fuel management began in 1978. The AECL (Atomic Energy of Canada Limited) was asked to develop the concept of spent nuclear fuel storage and the CCEA (*Commission de contrôle de l'énergie atomique*) initiated a regulatory research program focusing on granitic rock. In 1989, the Seaborn Commission was established to evaluate the results. The AECL builds on results from the Whiteshell Underground Laboratory in Manitoba. In 1994, the AECL submitted its Environmental Impact Statement (EIS). The CCEA considered that the storage of spent nuclear fuel in the Canadian Shield (Precambrian geological massif) would be feasible.

In 1996 and 1997, the Seaborn Commission conducted public hearings in five provinces. The CCEA found that the concept proposed by the AECL was acceptable and that a site should be selected. In 1998, the Seaborn Commission submitted its report to the Government of Canada. In 2002, the Government of Canada established the Nuclear Fuel Waste Act and the waste producers established the Nuclear Waste Management Organization (NWMO). The NWMO was to study three options and implement the most appropriate one accepted by the government. In the meantime, the CCEA became the CNSN (*Commission canadienne de sécurité nucléaire*, Canadian Nuclear Safety Commission, CNSC) in 2000. In 2005, the NWMO recommended an Adaptive Phased Management (APM) approach for the long-term management of spent nuclear fuel. In 2014, the CNSC established the Independent Advisory Group. In 2015, the NWMO continued site selection from nine candidates, all from the province of Ontario [CCS 17]. Over the past decade, Canada has implemented an "adaptive phased management" approach to dealing with its radioactive waste, involving storage at reactor sites and eventual final disposal in a centralized deep geological repository. The latter project will potentially take more than 140 years to fully implement.

Canada is developing several concepts for dry interim storage. One of the solutions adopted consists of dual-purpose (transport and storage) concrete containers with two steel inner and outer shells. Each OPG drum weighs 75 tons with 8.8 tons of fuel [IRS 18a].

Canada stores its spent fuel dry in casemates. The vertical-loading concept is the MACSTOR (Modular Air-Cooled STORage) from AECL (Atomic Energy Canada Limited). The AECL is also developing concrete containers that can hold from 1.3 to 10.3 tons of fuel with a thermal output of 1 to 2 kW. They are used for fuel for the Point Lepreau and Whiteshell CANDU reactors. A final concept, which has been in service since 1953 at the Chalk River site in Canada (WMA "B" facility), is used for experimental reactor fuels. This installation is a network of concrete tubes sealed on a concrete plate. Unfortunately, corrosion is strong and the spent fuel must be taken back and reconditioned [IRS 18a].

4.5.3.3. *Argentina*

In Argentina, nuclear waste management remains an unresolved issue. The *Commission nationale de l'énergie atomique* (CNEA) submitted the Strategic Plan of the National Radioactive Waste Management Program (*Plan stratégique du*

Programme national de gestion des déchets radioactifs, PEGRR) to the National Executive in 2014 for evaluation and approval [JIM 18]. The type of dry storage employed at Embalse is an older type developed at Chalk River in Canada (WMA "B" facility).

4.5.3.4. *Switzerland*

Switzerland is in the middle of implementing a very innovative approach to finding one or more storage sites. This approach includes provisions for transparency and public participation. To do this, Switzerland had to change its legal framework that limited the public's right to have a real say in the decisions to be made. This was achieved by abandoning the cantonal veto and granting the right to be heard on most issues.

The Swiss procedure for selecting sites for repositories for low- and intermediate-level waste and high-level waste consists of three stages, with the selection of potential siting regions, then the selection of at least two potential sites and finally the selection of the site and the start of the general authorization procedure. This procedure must follow the cantonal master plans and gather public input [AEN 10a]. Six areas are favorable for the establishment of a deep geological site [HOC 15].

Currently, Switzerland has a centralized storage facility in containers at Würenlingen [IRS 18a]. The cost of storing nuclear waste in Switzerland is estimated at between 14 and 25 billion Swiss francs (13.3 to 23.7 billion euros) depending on whether one accepts a cost certainty of 50 or 80% [BUD 18].

4.5.3.5. *Sweden*

Sweden began producing nuclear electricity in the early 1970s; today, the nuclear industry provides 50% of the country's total electricity production. Spent nuclear fuel from Swedish reactors is currently stored in SKB's Central Interim Storage Facility for Spent Nuclear Fuel (CLAB), on a peninsula where the Oskarshamn nuclear power plant is also located. This storage will be for 30 years in order to reduce radioactivity and heat emission. Currently, the spent fuel (about 4,000 tons in 2005, 6,000 to 8,000 tons in 2015) is stored at CLAB in large pools located at a depth of more than 30 meters. A new building will eventually allow 11,000 tons of nuclear fuel to be stored in the pools.

The project involves taking the fuel from CLAB and placing it in sealed copper containers with cast iron inserts. The containers will then be transported to a deep geological repository consisting of a system of horizontal tunnels at a depth of 400–700 meters in the granite. The tunnels will be about 250 meters long and spaced about 40 meters apart. Storage cavities will be located approximately every 6 meters on the floor of the tunnels. Copper containers will be deposited in the storage cavities and surrounded by a bentonite pad. Once deposited, the tunnels and shafts will be filled with a mixture of crushed rock and bentonite [MEA 07]. At CLAB, 6,700 tons of nuclear fuel are currently stored with a residual power of about 8.3 MW (Nystrom in [IAE 20]).

Three barriers are provided between the radioactive waste and the environment, the container, the bentonite plug and the granite rock. The container chosen by Sweden is special and designed to be particularly resistant. It is made up of two layers (container and over-container) and is nearly five meters long and a little over one meter in diameter. It weighs nearly 25 tons when full. The outer layer is made of thick copper and inside is a ductile iron (black steel) insert that provides sufficiently high mechanical strength (Figure 4.8). The bentonite pad has three functions in the repository, to prevent corrosive substances from reaching the container, to protect the container from minor movements and to delay any radionuclide that escapes as a result of a container leak. The granite in which the repository will be constructed is 1.8 billion years old and stable. Therefore, the probability of geological changes occurring in the next 100,000 years is negligible [MEA 07].

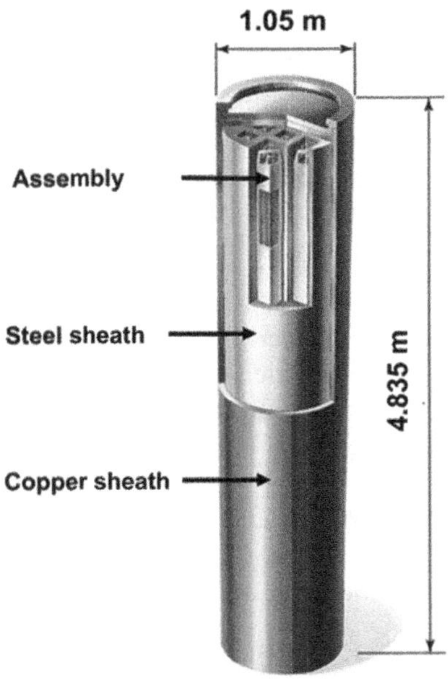

Approximate weight (kg):	
Fuel assemblies (BWR)	13,600
Steel container	3,600
Copper over-container	7,400
TOTAL	24,600

Figure 4.8. *Diagram of the copper container selected by Sweden to store spent fuel assemblies from BWRs (source: [ANO 20b]). For a color version of this figure, see www.iste.co.uk/amiard/radioactive.zip*

SKB's activities are financed by the Nuclear Waste Fund, which is fed by the payment of 0.01 SEK (~0.001 euros) per kWh consumed. It is thus the consumer who pays, and not the producer [MEA 07]. This financial model is strongly questioned by Kaberger and Swahn [KAB 15].

Research on nuclear waste management is largely carried out at the SKB underground laboratory in Äspö, north of Oskarshamn. The Äspö tunnel, 450 meters deep, is connected to the surface by access and ventilation shafts [MEA 07].

The Swedish site for geological storage of spent nuclear fuel was chosen in 2009 by SKB (Swedish Nuclear Waste and Spent Fuel Management Company). The site is located near the Forsmark nuclear power plant in the municipality of Östhammar [CNE 19a]. The application procedure for the deep geological repository for spent fuel began in March 2011. During the period 2011–2017, a formal analysis process was carried out with independent experts. The safety authority (SSM, *Strålsäkerhetsmyndigheten*), after a thorough review, gave a favorable opinion in January 2018 [CNE 18]. The Environmental Court, after six weeks of public hearings, recommended that the government request additional information from SKB on five specific aspects of the behavior of copper containers under storage conditions. The Environmental Court also gave a favorable opinion on the encapsulation plant and on the spent fuel storage site (CLAB) [CNE 18]. Since then, SKB prepared a response to this request and a package of documents was submitted to the government in early April 2019.

In parallel with the finalization of the licensing process, SKB is preparing for the start of construction of the storage facility in Forsmark and the encapsulation plant in Oskarshamn planned for 2022 and 2024, respectively, with the first commissioning operations around 2030 [CNE 19a].

4.5.3.6. *Finland*

About a quarter of the electricity consumed in Finland comes from the nuclear industry. The country has two nuclear power plants, Olkiluoto (owned by Teollisuuden Voima Oy, TVO) and Loviisa (owned by Fortum Power and Heat Oy), each with two reactors. In 2002, Finland decided to acquire a fifth reactor at Olkiluoto (OL3). The Finnish agencies responsible for nuclear power are the Nuclear Radiation and Safety Authority (STUK), which monitors safety issues, and Posiva Oy, the company responsible for the final disposal of spent nuclear fuel. It is owned by the two producing companies.

In addition to the three nuclear reactors (OL1 to OL3), the Olkiluoto nuclear site includes the interim storage of spent nuclear fuel, the underground repository for LLW, ILW and the ONKALO site (underground rock characterization facility). The spent nuclear fuel is stored in surface pools [MEA 07].

In 2013, Finland had 1,934 tons of spent fuel and 9,556 m^3 of LLW and ILW. Before the decommissioning of the six nuclear reactors, Finland had 11,500 tons of HLW [AUF 15].

For the final storage of spent fuel, four sites were examined, with regard to geological criteria, and the final choice fell on the municipality of Eurajoki (Olkiluoto peninsula), which voted in favor of storage in 2000. The construction of the underground storage should reach a depth of 500 m and began in 2017 [CNE 18]. The storage concept is based on the Swedish KBS3 technology with storage in crystalline rock of spent fuel encapsulated in copper containers. The world's first construction permit for a spent fuel repository was issued in 2015. Thus, geological disposal of high-level nuclear waste is becoming a reality in Finland. The Finnish government granted the construction license for Posiva on November 12, 2015, but this license was only valid for 2 years and the license reconfirms the retrievability requirement [VIR 17]. As of 2017, construction was well underway with approximately 7 km of tunnel already excavated. Drilling of the shaft to carry the encapsulated irradiated fuel containers into the storage facility is also under way [STU 17].

A full-scale mock-up test using electric heating elements to simulate the residual heat of spent fuel is being installed in a tunnel of the storage facility dedicated to observation in the framework of a multilateral cooperation project (called FISST, Fullscale In-situ System Test) [CNE 19a].

4.5.3.7. *Romania, Armenia and Czechoslovakia*

In Romania, storage at Cernavoda is in the form of dry storage in Canadian-designed, vertically loaded casemates.

Armenia is using dry casemate storage of Orano TN's NUHOMS horizontal loading system for its WWER reactor plant in Metzamor.

In the Czech Republic, seven sites for the deep geological repository have been pre-selected. The timetable foresees a final decision on the choice of site in 2025–2050 with the construction of an underground laboratory and the first research in it, the construction of a repository both above and below ground between 2050 and 2064, and commissioning in 2065 [BUR 15]. Dry storage in the form of a dual-use package (CASTOR process) has been adopted in the Czech Republic (WWER plant in Dukivany).

4.5.3.8. *Netherlands, Italy and Spain*

In the Netherlands, in 2012, the waste volumes for categories A, B, C and D were 1,450, 5,159, 2,686 and 1,000 m^3, respectively [ARE 15]. The centralized dry storage facility HABOG operated by COVRA at the Borsele site can accommodate a wide variety of nuclear waste [IRS 18a].

In Italy, in 2011, the volume of LLW was 22,200 m^3 and the volume of intermediate- and high-level waste (ILW/HLW) was 1,700 m^3 [DIN 15].

In Spain, the volumes of the first categories of waste (VLLW, LLW and ILW) are 176,000 m^3 and for the most hazardous waste (spent fuel, HLW and long-lived ILW), 12,800 m^3 [LOS 15]. Final storage in a deep geological layer is planned for the spent fuel elements. However, at this stage, there is no exploratory program for the selection of a site for such storage. The interim solution approved in 2011 is to construct a centralized interim storage facility, in addition to the three existing individual interim storage facilities at the operating nuclear power plant sites. The site for the centralized storage has already been selected and is located in Villar de Cañas, 140 km southeast of Madrid. The official authorization request for the project was sent to the Spanish Nuclear Safety Agency (CSN, *Consejo de Seguridad Nuclear*) by the national company in charge of nuclear waste management (ENRESA, *Empresa Nacional de Residuos Radiactivos*), but the Ministry of Ecological Transition asked the CSN at the end of 2018 to temporarily suspend the decision on the authorization of the project [CNE 19a]. This center will receive waste from the six active reactors and the two reactors being dismantled, amounting to more than 4,000 tons when it starts operating. This installation will eventually be able to receive 7,000 t, and its lifespan is planned for 60 years [AMI 13].

4.6. Conclusions

Long-lived intermediate- and high-level radioactive waste is particularly difficult to manage. Many solutions have been proposed for more than 50 years. The best solutions are those that combine a multitude of natural and artificial barriers. Two solutions meet these requirements: deep geological disposal and deep drilling. The majority of states have chosen the first solution, but no facility is operating at this time. The most advanced construction is in Finland.

Spent nuclear fuel, considered by some countries as nuclear material and not as waste, is one of the sources of high-level long-lived waste that requires rigorous management.

The IAEA and the NEA have made great efforts to inform and make recommendations to states that have to manage long-lived waste. The choice of site is essential and must be the subject of a consensus with the population. The roles of underground laboratories are extensively developed. Advice on the preparation of various dossiers (safety dossier, decision-making, long-term temporal evolution, post-closure monitoring) is detailed.

Public perception of the risk of high-level storage facilities is generally negative and results in opposition from local populations to siting in their vicinity. The authorities must deploy diplomatic resources and initiatives to ensure that the projects are accepted. Information and communication with the public must be careful, complete and sincere.

5

Nuclear Waste Management in France

5.1. Introduction

Historically, solid ILW and HLW was immersed at sea in black steel containers. France made little use of this method but did so until 1982. On the other hand, LLW generated during normal operation of the various nuclear installations (electronuclear reactors, research reactors, various nuclear cycle plants, etc.) is directly released into the environment (atmosphere, waterways and marine environment). Monitoring of environmental quality makes it possible to account for any excess of direct discharges in relation to regulatory authorizations for liquid and gaseous effluent discharges.

The French system is based on three essential and complementary pillars. Firstly, it includes a body of dedicated legislations and regulations. The second pillar is the National Radioactive Waste and Materials Management Plan (*Plan national de gestion des déchets et matières radioactifs*, PNGMDR). This is drawn up periodically by a multidisciplinary working group that meets several times a year and brings together waste producers, political and administrative representatives, the French National Agency for Radioactive Waste Management (*Agence nationale pour la gestion des déchets radioactifs*, ANDRA) and environmental protection associations. Finally, the last pillar is ANDRA, the agency dedicated to radioactive waste management, which was set up and thus plays a special role conferred on it by law [EVR 11].

France classifies its radioactive waste in five categories (see Chapter 1) [JOR 14]. The management of waste varies according to its category. All short-lived waste (TFAVC and FAVC) is managed by radioactive decay. Medium- and long-lived TFA waste is destined for surface storage and has recycling channels. For LLW and ILW (FA and MA), storage will be at the surface except for certain tritiated waste and certain sealed sources. A process is being studied in the context of Article 4 of the law of June 28, 2006, for the storage of LLW. Similarly, a method is being studied in the context of Article 3 of the law of June 28, 2006, for MAVL, HAVM and HAVL wastes.

The management chain for radioactive waste proposed by Boissier [BOI 11] is presented in Figure 5.1.

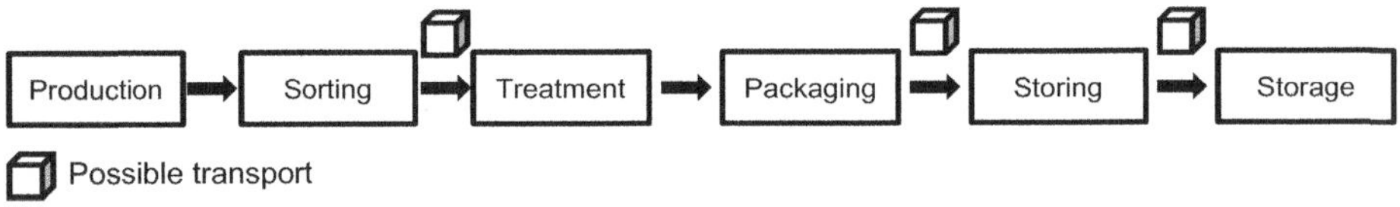

Figure 5.1. *Schematic representation of the radioactive waste management chain (source: modified from [BOI 11])*

ANDRA is responsible under the law of June 28, 2006, for publishing and updating the National Inventory of Radioactive Materials and Waste (*Inventaire national des matières et déchets radioactifs*) every three years. The latest inventory was published by ANDRA in 2018 [AND 18c] and takes into account waste until the end of 2016. In addition, an annual revision is published by ANDRA [AND 19, AND 20c].

France has some specificities in the management of its radioactive waste. Firstly, France has chosen to operate a closed nuclear fuel cycle. This has a strong impact on the quantities of radioactive waste, since new and spent fuels are considered as radioactive materials and not as waste. This situation is similar in a small number of countries (Russia, China, India, Japan). A second consequence of this choice is that the storage of all spent fuel is carried out in wet form and not dry, contrary to many countries such as the United States, where there are 71 dry storage facilities on site (NRC and Orano). This is necessary to facilitate reprocessing. Thus, France reprocessed from 2010 to 2016, 7,722 t of enriched natural uranium [HCT 18]. Nevertheless, local storage facilities are overcrowded, both near the reactors and at the centralized storage facility at La Hague, hence the need to plan new storage facilities.

The adoption of the closed fuel cycle implies nuclear reactors using MOX fuel and then a new generation of fast neutron reactors. The abandonment of this approach compromises the closure of the nuclear fuel cycle by France.

France also has the specificity, like the United Kingdom and Russia, of having large quantities of irradiated graphite waste whose final disposal has not been settled.

Several categories of radioactive waste do not yet have a dedicated channel in France. This is partly the case for long-lived intermediate- and high-level waste (ILW and HLW). French policy favors final disposal by deep geological burial. However, its Cigéo project is strongly contested, unlike in other countries such as Canada and Finland.

5.2. Direct discharges into the environment

The releases of radioactivity directly into the atmosphere and into the hydrosphere (rivers or ocean) from the French nuclear industry have not been as massive as the American, Soviet and British releases. This is partly due to the fact that the nuclear industry appeared in France later than in the three states mentioned above. A certain awareness that ionizing radiation could harm humans indirectly through exposure to their natural environment or through trophic transfer was beginning to be taken into account. The concept of dilution in environmental compartments, even though they are vast such as the atmosphere or the ocean, was therefore no longer favored. This type of release was limited to VLLW and LLW.

5.2.1. *The nuclear study centers*

In France, the CEA's nuclear study centers (Saclay, Fontenay-aux-Roses, Cadarache, Grenoble, etc.) release or have released many radionuclides directly into the environment. As an example, civil liquid (Table 5.1) and gaseous (Table 5.2) releases are provided for the year 2018, as well as releases related to the defense activities of this center for the same year (Table 5.3).

Installation	Tritium	Carbon-14	Beta emitters	Alpha transmitters
Civilian BNIs	$1.8.10^{-1}$	$1.7.10^{-2}$	$6.4.10^{-3}$	$6.7.10^{-5}$
ICPE	$7.9.10^{-1}$	$1.5.10^{-3}$	$3.9.10^{-1}$	$7.5.10^{-5}$

Table 5.1. *Liquid releases (in GBq) in 2018 at the Cadarache CEN (source: [CEA 19])*

Installation	Tritium	Rare gases	Iodine	Other β and γ emitters	Other α emitters
Civilian BNIs	$3.89.10^{1}$	$1.17.10^{4}$	$1.14.10^{3}$	$2.04.10^{-3}$	$1.19.10^{-5}$
ICPE	$1.15.10^{-1}$	0.00	$1.60.10^{-5}$	$2.33.10^{-5}$	$6.44.10^{-6}$

Table 5.2. *Gaseous releases (in GBq) in 2018 at the Cadarache CEN (source: [CEA 19])*

Effluents	Tritium	^{14}C	Rare gases	Iodine	Global beta	Total alpha
Liquids	$4.5.10^{-3}$	–	NC	$1.5.10^{-3}$	$2.7.10^{-4}$	$7.3.10^{-6}$
Gaseous	$3.99.10^{-2}$	$1.99.10^{-1}$	$6.60.10^{3}$	$2.84.10^{-3}$	$1.17.10^{-4}$	$1.62.10^{-5}$

Table 5.3. *Liquid releases (transferred to the STEP EI) and gaseous releases (in GBq) in 2018 from the INBS PN of the Cadarache CEN (source: [CEA 19])*

5.2.2. *Nuclear reactors*

In normal operations, all reactors have liquid discharges into rivers or the sea and gaseous discharges directly into the atmosphere. These discharges vary from one reactor to another, depending on its type and age. Table 5.4 gives the average liquid and gaseous discharges for the two main types of reactors currently operating in France (PWR 900 and 1300 MWe).

Liquid discharges			Gaseous discharges		
Reactor	**900 MWe**	**1300 MWe**	**Reactor**	**900 MWe**	**1300 MWe**
Iodine (GBq.yr^{-1})	0.009	0.008	**Noble gases (TBq.yr^{-1})**	0.9	2.2
^{14}C (GBq.yr^{-1})	10.6	16.2	**^{14}C (TBq.yr^{-1})**	0.14	0.22
Tritium (TBq.yr^{-1})	10.2	24.1	**Tritium (TBq.yr^{-1})**	0.22	1.18
Other (GBq.yr^{-1})	0.5	0.6	**Iodine (GBq.yr^{-1})**	0.024	0.040
			Other (GBq.yr^{-1})	0.003	0.004

Table 5.4. *Liquid and gaseous discharges from the principal radionuclides from two of the three types of French reactors (PWR of 900 and 1300 MWe) in 2003 (source: [EDF 04] EDF (2004), in [AMI 13])*

5.2.3. *Fuel cycle plants*

All fuel cycle plants have direct releases into the environment. The most significant ones are from reprocessing plants. In France, the La Hague plant releases large quantities of radioactive inert gases (xenon, krypton) and tritium into the atmosphere (Table 5.5).

Annual liquid discharges (2002)		Annual limits	Annual gaseous discharges (2002)		Annual limits
Tritium (TBq)	11,000	18,500	**Tritium (TBq)**	63.2	150
^{137}Cs and ^{90}Sr (TBq)	1.42	3.2	**Halogens (GBq)***	5.42	18
Alpha emitters (TBq)	0.039	0.07	**Aerosols (MBq)****	109	
Other (TBq)	23.3	30	**Other (TBq)*****	245,000	470,000

Table 5.5. *Liquid and gaseous releases for 2002 from the La Hague reprocessing plant (source: [SEY 11] in [AMI 13]) and annual regulatory limits (source: [JOR 16]). *Especially ^{129}I; ** especially ^{106}Ru; *** especially ^{85}Kr and ^{14}C*

5.3. The inventory of nuclear waste in France

France has two main types of radioactive waste, military waste and civilian waste. ANDRA is in charge of their inventories. The inventory of civilian waste is updated every three years, with a lighter update every year.

5.3.1. *Military waste*

For a long time, French military waste was protected by defense security and its location and quantities were not officially known, nor precisely. One of the first pieces of information on military radioactive waste was provided by the parliamentary report by Bataille [BAT 97]. This situation has improved with the TSN law of 2006. However, the national inventory established by ANDRA cannot detail the location or the composition of the fuels of reactors on board aircraft carriers and submarines, which are in a sensitive sector. These fuels represent a small part of the fuels generated by civilian reactors. The defense spent fuel stockpile was estimated at 146 tons in 2010 and was 194 tons at the end of 2018 [AND 20c].

The management of military radioactive waste depends on either the Ministry of Defense or the Ministry of Energy. Among these wastes, some are managed on a military level, others on a civilian level and sometimes in a mixed way.

There are 22 BNIs under the authority of the Minister of Defense. Radioactive waste is found in the three naval arsenals (Ile Longue, Toulon and Cherbourg). The arsenals produce waste because of the maintenance operations of the nuclear reactors on board for the propulsion of submarines and the French aircraft carrier. The reformed reactor cores are stored in Cherbourg. According to the ASND, waste from military ports remains only for a short time on the sites themselves. It is evacuated either to ANDRA storage facilities when the channels exist, or to CEA sites for storage.

The test centers of Bourges and Gramat carry out experiments and tests on weapons containing uranium in a form depleted in the isotope-235 [AMI 13].

Apart from these sites under the Ministry of Defense, there are seven BNIs in France that come under the Ministry of Energy. Among these sites, some are mixed, with the presence on the same site of a civilian BNI (whose safety authority is the ASN) and other sites that are purely BNIs (whose competent safety authority is the DSND alone). The mixed sites jointly managed with the CEA/DEN are Cadarache and Marcoule, and the one managed with Areva-NC is Pierrelatte. The sites of Valduc, DAM Ile-de-France and CESTA are BNIs dependent only on CEA/DAM, while the SODERN site is a BNI managed by EADS/Sodern [DSN 10].

The principal military wastes are stored in the BNIs of Pierrelatte, Marcoule and Valduc (Table 5.6). The wastes of the CEA's *Direction des applications militaires* (DAM) have two particularities: the quantities produced are small compared to civilian production and they are almost exclusively contaminated by alpha emitters or tritium. The long-lived intermediate-level wastes (LL-ILW) are essentially sludges and concentrates, formerly produced by the Valduc effluent treatment station, which have been injected into metal drums [AMI 13].

Site	VLLW	SL-ILW	LL-LLW	LL-ILW
Long Island	34	15	1	
Toulon	66	24	3	
Cherbourg	490	27		
Marcoule	16,000	59,900	2,200	14,800
Pierrelatte	27,000	440		
Cadarache	322	280		
Valduc (excluding tritium)	9,000	425	100	
DAM Ile de France	500	165	11	
Total	**53,412**	**61,276**	**2,315**	**14,800**

Table 5.6. *Radioactive waste of various categories (volumes expressed in m^3) (source: [DSN 10])*

At Marcoule, the site's old wastes essentially come from reprocessing operations on irradiated UNGG fuel, in order to recover the plutonium. These operations were carried out in the various workshops of the UPI plant and at the Marcoule pilot workshop (*atelier pilote de Marcoule*, APM). The great majority of the waste from these activities was conditioned: in the Marcoule vitrification workshop (*atelier de vitrification de Marcoule*, AVM) for HLW and in the bitumen workshop of the *Station de Traitement des Effluents Liquides* (STEL) for ILW[1]. The conditioned vitrified solid waste represents about 580 m^3 of vitrified HA waste or waste to be vitrified (fission products, U and Pu). This old waste, for that which has a disposal route, is taken up and reconditioned by the CEA, and for the others an improvement in the conditions of storage of the waste while waiting for an outlet [DSN 10]. The latest assessments by ANDRA [AND 18c] list 3,159 vitrified HA containers, or 553 m^3, and 314 vitrified containers of technological waste classified as LL-ILW, or 55 m^3, for a total of 608 m^3.

At Pierrelatte, unconditioned technological waste or contaminated rubble, contaminated by uranium, is buried under mounds. This waste, most of it VLLW of CEA origin, essentially comes from the former gaseous diffusion plants (*usines de diffusion gazeuse*, UDG).

At Valduc, there are two main categories of waste: waste contaminated with alpha emitters (known as "alpha waste") and tritiated waste. The wastes that have not yet been disposed of are stored in supervised installations, either at the CEA center at Cadarache (LL-ILW alpha waste) or at the Valduc center (tritiated waste), awaiting storage solutions. In addition to the waste contaminated by alpha emitters listed in Table 5.6, the Valduc center encloses solid waste contaminated by tritium and conditioned. These are 2,000 m^3 of tritiated SL-ILW, 150 m^3 of tritiated SL-ILW contaminated by uranium and 30 m^3 of tritiated VLLW [DSN 10].

Military waste from nuclear testing in the Pacific presents a special case. Much of this military waste was buried in boreholes. These wells were drilled to a depth of about 1,200 m in the Mururoa atoll. They are primarily intended for the burial of α waste. They have larger diameters than the firing pits in order to increase storage capacity: a diameter of about 2 m for PS1 and about 1.8 m for PS3. Bulk aggregates and 100 and 225 liter concrete drums were buried per section with cement plugs in PS3 (large diameter borehole). Well PS1 contains 21,500 TBq and well PS3 377 TBq, mostly alpha emitters for both wells. The containment of radionuclides is ensured, over time, by three successive barriers, trapping in the lava, retention on the

1 Packages of bituminous alpha waste pose a huge problem because they give off hydrogen and the risk of fire is too great for them to be accepted in the Cigéo storage center. They must be reconditioned to become chemically inert. Studies are under way at the CEA, supervised by the ASN (see Chapter 2).

volcanic rocks, particularly in the "chimney-cavity" and containment of pore water, the vehicle for radionuclides in the geological formations. The residual radionuclides likely to have a long-term impact on the biosphere are ^{239}Pu and ^{241}Am and, in the medium term, ^{90}Sr and ^{137}Cs. Their blockage in the massif is therefore linked to the quality and durability of the successive barriers.

5.3.2. *Civilian waste*

The latest civilian waste inventory conducted by ANDRA dates from 2018 [AND 18c] and takes into account the radioactive waste inventories at the end of 2016. The volumes of waste in the six categories are listed in Table 5.7, along with the activity levels (Table 5.8). A significant portion (Table 5.9) is stored at the producers' own sites.

Category	Volume at the end of 2016 (m^3)/end of 2018 (m^3)	Percentage (%)
HLW	3,650/3,880	0.2
LL-ILW	45,000/43,000	2.9
LL-LLW	90,500/93,700	5.9
SL-ILW	917,000/945,000	59.6
VLLW	482,000/557,000	31.3
WWC*	1,800/1,350	<0.1
Total	1,540,000/1,640,000	100.0

Table 5.7. *Waste in various categories by volume (m^3) and percentage of total at the end of 2016 (source: [AND 18c]). End of 2018 *WWC: waste without a channel*

Category	Activity at the end of 2016 (TBq)	Percentage (%)
HLW	194,000,000	94.9
LL-ILW	10,100,000	4.9
LL-LLW	280,000	0.14
SL-ILW	52,000	0.03
VLLW	300	0.0001
Total	**205,000,000**	**100.0**

Table 5.8. *Radioactivity levels of various categories of radioactive waste in TBq or 10^{12} Bq^3) and as a percentage of total at the end of 2016 (source: [AND 18c])*

Category	Volume at the end of 2016 (m^3)
HLW	3,650
LL-ILW	45,000
LL-LLw	90,500
SL-ILW	74,100
VLLW	154,000
WWC*	1,800

Table 5.9. *Waste of various categories by volume (m^3) stored at waste generator sites at the end of 2016 (source: [AND 18c]). *WWC: waste without a channel*

The largest quantities of waste come from nuclear power generation, followed by nuclear research (Tables 5.10 and 5.11). The total waste masses are from both the nuclear and non-nuclear industries (Table 5.12).

Electronuclear	Research	Defense	Non-nuclear industry	Medical
58.8	27.7	9.4	3.6	0.6

Table 5.10. *Distribution of radioactive waste by sector at the end of 2016 (in percent) (source: [AND 18c])*

Volume (m^3)	Electronuclear	Research	Defense	Non-electronuclear industry	Medical
HLW	3,250	161	232	-	-
LL-ILW	27,900	10,700	6,300	161	2
LL-LLW	38,300	13,900	18,000	20,300	-
SL-ILW	592,000	232,000	63,100	22,200	8,410
VLLW	243,000	170,000	56,500	12,100	88
Total	**905,000**	**427,000**	**144,000**	**54,700**	**8,500**

Table 5.11. *Distribution of the total volume (m^3) of waste by economic sector and category at the end of 2016 (source: [AND 18c])*

For the various sectors of activity (electronuclear, research, defense, non-nuclear industry), the total volume of waste comes from the nuclear and non-nuclear industries (Tables 5.13 and 5.15–5.17). The volumes of specific waste from Orano's Malvési plant are significant (Table 5.14).

Economic sector	Quantity at the end of 2016 (in tML)
Electronuclear	395,000
Research	210
Defense	177 tons
Non-electronuclear industry	6,400
Medical	-

Table 5.12. *Distribution of the total mass of radioactive material by economic sector at the end of 2016 expressed in tons of heavy metal (tML, except for defense in tons) (source: [AND 18c])*

Category	Volume (m^3) at the end of 2016	Percentage (%)
HLW	3,250	0.4
LL-ILW	27,900	3.1
LL-LLW	38,300	4.2
SL-ILW	592,000	65.4
VLLW	243,000	26.9
Total	**905,000**	**100.0**

Table 5.13. *Summary of the radioactive waste from the nuclear power sector (source: [AND 18c])*

Waste from Malvési	Volume (m^3) at the end of 2016
Settling tank sludge	70,400
Historical RICU	282,000
Nitrated effluents	374,000

Table 5.14. *Volume of waste from Orano's Malvési plant (source: [AND 18c])*

Category	Volume (m^3) at the end of 2016	Percentage (%)
HLW	161	-
LL-ILW	10,700	2.5
LL-LLW	13,900	3.3
SL-ILW	232,000	54.3
VLLW	170,000	39.9
Total	**427,000**	**100.0**

Table 5.15. *Summary of research sector radioactive waste volumes at the end of 2016 (source: [AND 18c])*

Category	Volume (m^3) at the end of 2016	Percentage (%)
HLW	232	0.2
LL-ILW	6,300	4.4
LL-LLW	18,000	12.5
SL-ILW	63,100	43.7
VLLW	56,500	39.2
Total	**144,000**	**100.0**

Table 5.16. *Summary of defense sector radioactive waste volumes at the end of 2016 (source: [AND 18c])*

Category	Volume (m^3) at the end of 2016	Percentage (%)
HLW	-	0.0
LL-ILW	161	0.3
LL-LLW	20,300	37.1
SL-ILW	22,200	40.6
VLLW	12,100	22.0
Total	**54,700**	**100.0**

Table 5.17. *Summary of radioactive waste volumes from the non-electricity industry sector at the end of 2016 (source: [AND 18c])*

After the 2018 inventory, ANDRA updated the waste productions for calendar years 2017 and 2018 in 2019 and 2020 (Table 5.18).

	Annual change (m^3)	
Category	2017	2018
HLW	+90	+140
LL-ILW	-2,800	+200
LL-LLW	+3,100	+100
SL-ILW	+21,000	+6,000
VLLW	+55,000	+20,000
WWC	-30	-420
Total	**+80,000**	**+30,000**

Table 5.18. *Annual change in the volume generation (m^3) of waste in 2017 and 2018 (source: [AND 19, AND 20c])*

It should be stated that ANDRA does not currently store HLW, LL-ILW and LL-LLW. The summaries concern wastes stored at the producers' sites.

Inventories of all radioactive materials (natural uranium, uranium from reprocessing, enriched or unenriched uranium oxide fuel, mixed uranium and plutonium fuel) increased during 2017 and 2018. Only research reactor fuel, defense spent fuel and thorium in stockpiles remain stable [AND 19, AND 20c]. On the other hand, the quantities of French plutonium increase each year. For example, they increased from 37.7 t in 2010 to 44.1 t in 2016, to which can be added the quantities of foreign plutonium. Thus, in 2016, there were 14.9 t of plutonium belonging to Japan [ZER 18b].

To estimate radioactive waste stocks in the future (Table 5.19), ANDRA simulated four scenarios. Scenario SR1 assumes the continuation of nuclear power generation with the deployment of EPR™ and then RNR reactors and the continuation of spent fuel reprocessing (continuation of the current strategy). The operating times of the current reactor fleet are increased to 50 and 60 years. The SR2 scenario is identical to the SR1 scenario, but with a uniform 50-year operation of all the reactors. The SR3 scenario is based on the continuation of nuclear power generation with the deployment of EPR™ reactors only. The SNR scenario assumes the non-renewal of the existing fleet, resulting in the immediate cessation of nuclear power.

	Volumes of radioactive waste at completion (m^3)			
Category	**SR1**	**SR2**	**SR3**	**SNR**
HLW	12,000	10,000	9,400	4,200
LL-ILW	72,000	72,000	70,000	61,000
LL-LLW	190,000	190,000	190,000	190,000
SL-ILW	2,000,000	1,900,000	2,000,000	1,800,000
VLLW	2,300,000	2,200,000	2,300,000	2,100,000
Radioactive materials requalified as radioactive waste at completion (tML)				
Natural uranium			470,000	434,000
Spent fuel (ENU, ERU)			3,700	25,000
Spent fuel (MOX, RNR)			5,690	3,590
Other fuels			5	54
Plutonium			-	2
Other materials			70	70

Table 5.19. *Quantities of radioactive waste at completion estimated in volumes (m^3) and of radioactive material reclassified as radioactive waste expressed in tons of heavy metal (tML) according to the four scenarios retained (source: [AND 20c])*

5.4. Nuclear waste management in France

The management of nuclear materials and radioactive waste in France is based on several laws. The creation and functions of the various official organizations working in this field are specified. Similarly, the roles of the many actors involved in this management are defined.

5.4.1. *The regulatory context*

In France, two laws and their implementing regulations concerning radioactive waste management were passed in 1991 and 2006, respectively, with Christian Bataille (PS, Nord) and Claude Birraux (UMP, Haute-Savoie) as rapporteurs. These laws set the rules of the game in this area and gave a pre-eminent role to the French National Agency for Radioactive Waste Management (*Agence nationale pour la gestion des déchets radioactifs*, ANDRA).

On December 30, 1991, France passed a law called the "Bataille Law", named after the deputy who was its main promoter, which organizes the future management of waste. This law provides for a major research program in three areas. The first is to explore the possibility of separating and transmuting radioactive elements. The second is the study of deep geological disposal of long-lived radionuclides. The third axis aims to ensure the conditioning of radioactive elements and their storage, on the surface, over a long period [JOR 92a].

The second law is the law of June 13, 2006 on transparency and security in nuclear matters, which created the *Autorité de sûreté nucléaire* (ASN), an independent administrative authority, and gave it powers to control the safety of basic nuclear installations, a status that includes radioactive waste management facilities. This law also includes provisions relating to public information in the field of nuclear safety [JOR 06a].

After 15 years of research carried out in accordance with the 1991 law, the ASN has issued several opinions, summarizing various scientific, political and public reports. Concerning research axis 1, the ASN considers that the technological feasibility of separation and transmutation has not yet been established [ASN 13]. Even though such a solution were implemented, the elimination of high-level and long-lived radioactive waste would not be complete. Another key solution is necessary. With regard to axis 3, the ASN considers that long-term storage cannot be a definitive solution for the management of high-level, long-lived radioactive waste. The ASN therefore considers that deep geological disposal is a definitive management solution that cannot be ignored. The ASN believes that it is desirable to manage the repository in stages, from the start of operation of the repository to its closure. The decision to close the disposal facility, and thus to end reversibility, should be made by Parliament [ASN 06].

The third law is that of June 28, 2006 [JOR 06b]. It concerns the program for the sustainable management of radioactive materials and waste. It defines the national policy for the sustainable management of radioactive materials and waste. This law organizes the sustainable management of radioactive materials and waste and its financing. This law introduces into the legislative framework new provisions relating, in particular, to the preparation of a national plan for the management of radioactive materials and waste, the more precise definition of ANDRA's missions and the assessment by the operators of basic nuclear installations of the costs of managing their spent fuel and radioactive waste, as well as the constitution of provisions relating to these costs. Its main articles have been the subject of application decrees [CHE 11].

In addition, the 2006 law establishes a precise schedule for waste management. In 2012, the CEA had to specify the industrial perspectives of separation and transmutation. In 2015, new storage facilities were proposed. In the same year, a request for authorization for a reversible deep geological repository had to be examined and a new law determined the conditions for reversibility. In 2025, the operation of this possible underground storage facility must be planned [AND 06b].

Other French regulations exist, such as the one governing sealed sources [JOR 15a] or the compliance between the PNGMDR and the environmental code [JOR 15b].

At the European Union level, Council Directive 2011/70/EURATOM of July 19, 2011 published in the *Journal officiel* of the European Union of August 2, 2011 [JOU 11] establishes a community framework for the responsible and safe management of spent fuel and radioactive waste.

5.4.2. *The National Radioactive Materials and Waste Management Plan (PNGMDR)*

The law of June 28, 2006 created the National Radioactive Materials and Waste Management Plan (*Plan national de gestion des matières et déchets radioactifs,* PNGMDR). This plan is designed to find solutions that guarantee transparent, rigorous and safe long-term management of all radioactive waste in France, regardless of its origin. The PNGMDR is a strategic tool for the management of radioactive materials and waste. Its objective is to assess the existing management methods for radioactive materials and waste, to identify the foreseeable needs for storage facilities, and to specify the capacities required for these facilities and the storage periods. For radioactive waste that has not yet been definitively managed, the plan determines the objectives to be achieved. The National Plan also organizes the implementation of research and studies on the management of radioactive materials and waste, setting deadlines for the implementation of new management methods, the creation of facilities or the modification of existing facilities [EVR 11].

The governance of the PNGMDR is carried out at several levels. The national management plan is prepared by the DGEC (*Direction générale de l'énergie et du climat*, General Directorate for Energy and Climate) of the Ministry of Ecological Transition and the ASN. It mobilizes several stakeholders at different stages. The assessment of research is carried out by the *Commission nationale d'évaluation*, CNE2, National Evaluation Commission, the preparation of measures is carried out by the PNGMDR working group, the assessment of the impact of the plan gives rise to an opinion from the *Autorité environnementale* (Ae. Environmental Authority) and the effectiveness of the system is assessed in an opinion from the *Office parlementaire d'évaluation des choix scientifiques et technologiques* (OPESCST, Parliamentary Office for the Evaluation of Scientific and Technological Choices) and, since 2016, the public has given its opinion during the preparatory debate on the updating of the PNGMDR [CND 19a].

The PNGMDR is revised every three years. The fifth PNGMDR (2019–2021) defines the medium-term objectives (2030) and they mainly concern TFA waste from dismantling, LL-LLW and the fate of radioactive materials, as well as the Cigéo project.

Originally, three research avenues, also called "axes", were chosen by the law of December 30, 1991, concerning the future of high-level, long-lived radioactive waste. Axis 1 concerned separation–transmutation, axis 2 coordinated research on deep geological disposal and axis 3 supervised research on conditioning and long-term surface storage. The CEA was in charge of research for axes 1 and 3 and ANDRA for axis 2. The producers of spent fuel and radioactive waste are responsible for these substances [IRS 13c].

5.4.3. *The different actors in nuclear waste management in France*

The PNGMDR sets out the major orientations for radioactive waste management. Its progress is regularly assessed by the administrative, political, technical-scientific and public spheres, which send their comments and recommendations to the waste-generating operators.

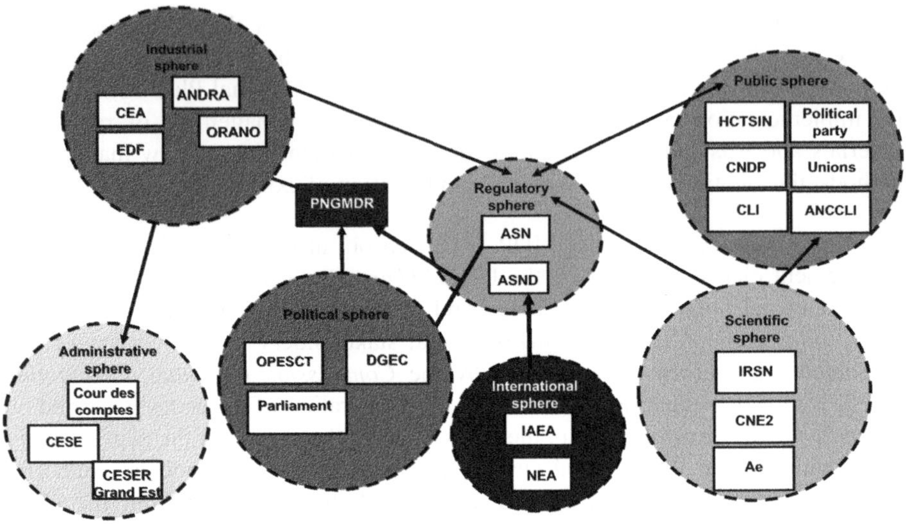

Figure 5.2. *The main actors and the various spheres involved in radioactive waste management in France. All the interactions between the various actors are not shown in order not to make the picture too crowded. For a color version of this figure, see www.iste.co.uk/amiard/radioactive.zip*

5.4.3.1. *The National Agency for Radioactive Waste Management (ANDRA)*

ANDRA is the contracting authority for the management of French radioactive waste. Originally, ANDRA was simply an agency of the CEA [JOR 69][2]. In France, the environmental code states that ANDRA, a public industrial and commercial establishment, is responsible for long-term management of radioactive waste, and, in particular, for the inventory of radioactive materials and waste present on national territory, for carrying out research and studies on deep geological storage and disposal, and for defining specifications for the storage of radioactive waste.

5.4.3.2. *The main producers of radioactive waste*

The CEA is involved in the PNGMDR at two levels, as the leader of research areas 1 and 3 and as a producer of radioactive waste through the various nuclear facilities it operates. In France, the other two main producers of radioactive waste are EDF and Orano.

2 This decree of June 1969, with only its title and no explanatory text, shows that at that time the political authorities were not aware of the importance of the safety of radioactive waste.

5.4.3.3. *The official circuit of the administrative and political spheres*

The French Nuclear Safety Authority (*Autorité de sûreté nucléaire*, ASN) is the national regulator for nuclear waste safety.

The Government and Parliament legislate and enact laws concerning radioactive waste management, regulating the decision-making processes and creating the bodies involved in this management. The DGEC (*Direction générale de l'énergie et du climat*) prepares the PNGMDR, which is updated every three years. This plan details the schedule for the files and research to be carried out. These will be carried out mainly by ANDRA and the CEA.

Each year, the Government sends Parliament a report on the progress of research, mainly from ANDRA and CEA. Parliament refers the report to the *Office parlementaire d'évaluation des choix scientifiques et technologiques* (OPECST). This Office is made up of deputies and senators, representatives of the Assemblée national and the Sénat.

The implementation of the PNGMDR is also examined each year by various bodies, each with its own competence. These are the *Commission nationale d'évaluation* (CNE2), the *Autorité environnementale* (Ae), the *Conseil économique, social et environnemental* (CESE) and its most involved regional chamber (CESER Grand Est) and the *Cour des comptes*.

Thus, the CESER believes that the issue of radioactive materials and waste management should not be confined to a debate among experts. It is first and foremost an ethical issue, and it is up to society to address it. On the one hand, Cigéo, if it is well designed and constructed in accordance with the principle of reversibility, can offer a safe and ultimately sustainable solution that would not require active monitoring of HLW and LL-ILW for centuries. On the other hand, the continuation of experiments on long-term storage would make it possible to recreate the possibility of a choice. If this solution were adopted, it would provide at least a century for scientific research to try to reduce the harmfulness and longevity of radioactive waste. However, the final decision to begin storage in Cigéo is not an emergency because the site will not be able to receive the first packages of HLW and LL-ILW for several decades [FAU 19]. On the other hand, irradiated fuel only needs to cool for 60 years.

5.4.3.4. *The technical and scientific circuit of the scientific sphere*

The ASN requests a technical opinion from IRSN and the *Groupe Permanent Déchets* (GDE) for each dossier submitted by ANDRA. The opinions are posted on the ASN and IRSN websites.

The 1991 Act established an independent National Evaluation Commission (*Commission nationale d'évaluation indépendante*, CNE). It was established in 1991 and is composed of 12 members appointed for their expertise in the field. It published 11 reports from 1995 to 2005 and a comprehensive report in 2006. They are all on its website. It was replaced in 2006 (law of June 28, 2006) by the *Commission nationale d'évaluation des recherches et études relatives à la gestion des matières et des déchets radioactifs* (CNE2). It has 12 members and 3 invited members. Since 2007, it has published 13 reports, also on its website. The CNE2 has also produced analyses of reversibility, the Cigéo project, separation–transmutation, storage, radioactive materials and waste, and bitumen. These last three reports date from 2018, 2019 and 2020 [CNE 18, CNE 19a, CNE 20]. In particular, the CNE has established an international overview of HLW management in which it summarizes the experiences of Finland, Sweden, Belgium, English speaking Canada and Spain. In its latest report, the CNE questions the consequences of abandoning the fast reactor project for at least 30 years on the backend of the fuel cycle. It emphasizes that the use of MOX for existing reactors, multi-recycling in PWRs, the development of new reactors, the manufacture of new fuels and the treatment of spent fuel are highly interdependent. The CNE is concerned about the lack of a clear vision for the management of spent fuel, which could reach 250,000 m^3. It also recommends that the institutional players concerned establish the necessary consultation to deal with the many and sometimes contradictory issues at stake. The CNE notes that the time table for the Cigéo project is drifting steadily, even though it recognizes that ANDRA has made substantial progress and has reached scientific and technical maturity. The CNE regrets that France has little knowledge of the behavior of traditional repositories, even though they represent only a wait-and-see solution, and recommends a proactive policy for developing a permanent storage solution.

5.4.3.5. *The public circuit*

With the publication of the TSN law, the *Haut comité pour la transparence et l'information sur la sécurité nucléaire* (HCTISN) was created. This body regularly produces reports and summaries in the nuclear field on subjects of interest to the public.

At the level of each nuclear installation, a Local Information Committee (*Comité local d'information*, CLI) is installed with political representatives and the public. Some of these CLIs are grouped in a national association (ANCCLI, *Association national des comités et des commissions locales d'information*).

Each new PNGMDR is the subject of public debates scheduled by the National Commission for Public Debate (*Commission nationale du débat public*, CNDP). These public debates are organized by the Commission (CPDP). The CLI and

ANCCLI, the defense associations, the unions and the operators actively participate in these debates, giving their opinions and recommendations and writing briefs on various subjects.

The last public debate took place from April 17 to September 25, 2019 with 23 meetings in Paris and the provinces. Special methods of participation were inaugurated with a process of clarification of technical controversies, a mirror group, a workshop for the next generation, mobile debates and online expression. This debate was also original in that it dealt with a plan and not a project, and that the complex technical aspects of the topics to be discussed were numerous and could lead to it being a debate of experts only. The non-specialized public was less numerous than desired, and few sessions were disrupted by opponents of the principle of the debate itself.

The preparation of the next National Management Plan, which was the reason for the CNDP's referral, called for discussion of five questions posed in the project owner's dossier: the recovery of substances classified as "materials" and not as "waste", storage capacities for spent fuel, the treatment of large volumes of VLLW, the management of LL-LLW, and the industrial pilot phase and reversibility of the deep geological disposal project (Cigéo project) [CND 19a].

The challenges to the PNGMDR are still strong and numerous. They are relayed by the public debate, notably in the 62 stakeholder reports and 22 contributions [CND 19b]. These papers and contributions are very varied and come from environmental protection associations, industrial operators, unions, political parties, experts and official bodies.

Many of these interventions concern VLLW, with those in favor of adopting clearance levels and those violently opposed. Others concern the Cigéo project, with clear-cut positions for and against radioactive materials and their actual classification as waste. Several examples will be detailed in section 5.8.

5.4.3.6. *The international circuit*

The Nuclear Energy Agency (NEA) of the OECD has issued several opinions concerning French nuclear waste management projects. Its opinion on the ANDRA files, Argile 2001 and Argile 2005 [AEN 03, AEN 06b], was relatively favorable. The Agency concluded that ANDRA's scientific and technical program was consistent with international best practices and in several areas at the forefront of waste management programs. It considers that the overall strategy for long-term safety developed by ANDRA is reliable and comprehensive. It considers that the design developed by ANDRA satisfies the requirement to prove the principle of reversibility and that this is not achieved at the expense of long-term safety. It positively notes the significant improvement in confidence in the high performance of the COX (Callovo-Oxfordian) as a barrier to the migration of radionuclides and other chemicals into the surrounding environment and the biosphere.

The ASN submitted the safety option file for the future industrial geological disposal center (CIGEO) to the IAEA in order to benefit from the knowledge as well as the feedback of international experts who have participated in comparable projects. To this end, a peer review mission, led by the IAEA, was requested by France from the Department of Nuclear Safety and Security, which took place from November 7 to 15, 2016. The experts examined the dossier submitted by ANDRA with regard to the safety standards established by the IAEA. The experts' examination specifically focused on three aspects: the R&D program in connection with the development of the project, the monitoring of CIGEO as planned by ANDRA, and the definition of scenarios for both operational and long-term safety.

The IAEA experts have reasonable confidence in the robustness of the deep disposal concept and in ANDRA's ability to design a safety demonstration. They have identified some post-closure scenarios and aspects of the operational design that ANDRA should take into account in order to further strengthen confidence in the safety demonstration. They also suggested improvements in ANDRA's research and development plan and in the monitoring of the development program [IAE 16c].

5.5. The organization of storage for identified waste

France classifies its waste into six categories: VTC, TFA, FMA-VC, FA-VL, MA-VL and HA [JOR 14]. However, France manages its radioactive waste according to four channels because the categories of FA-VC and MA-VC, as well as MA-VL and HA-VL, are grouped together. France differs from the majority of other States in not accepting any release threshold.

5.5.1. *The various types of containers*

For low-activity and VLLW packages, containers and conditioning matrices are sufficient. The containers are made of concrete, unalloyed steel (ordinary steel) or alloyed steel (stainless steel).

5.5.1.1. *Packages for transport*

In France, there are four classes of packages for transport. Type A packages for LLW, B packages for HLW (> 50 mSv at 1 m in 30 minutes) that must withstand accident situations, C packages for air transport and IP packages for industrial transport. In 2012, the CNRS launched a federative research project called Needs

(*Nucléaire: énergie, environnement, déchets, société*, literally translated as Nuclear: energy, environment, waste, society) that focuses on the treatment and packaging of radioactive waste.

5.5.1.2. *Conditioning of long-lived intermediate-level waste*

LL-ILW is mainly composed of metal structures. In France, between 1990 and 1995, this waste was packaged in concrete. Their average activity is 50 TBq for beta emitters and 0.1 TBq for alpha emitters These drums have a dose rate of about 10 $Gy.h^{-1}$ on contact. At the end of 2018, ANDRA estimated that 1,518 packages of cemented cladding and connectors were stored at La Hague, representing 2,277 m^3 of waste classified as LL-ILW [AND 18c].

Since 2000, the cementing process has been replaced by a process of compacting the liner sections, which allows a reduction in volume of about a factor of 5 compared to the cementing of these B wastes likely to go into deep storage. The wafers thus formed are stacked in a stainless steel container (5–10 wafers per package). The package thus formed is called a Standard Compacted Waste Package (*Colis Standard de Déchets Compactés*, CSD-C). This package, whose initial activity is of the order of a few hundred $TBq.g^{-1}$, also contains alpha emitters [CEA 08].

5.5.1.3. *Conditioning of high-level waste in France*

France has chosen vitrification to confine its HLW. The containment process takes place in three stages: evaporation of the fuel dissolution solutions, calcination (100–300 C) and vitrification (1,050–1,300 C). The first prototype processes were GULLIVER (1964–1967) and PIVER (1968–1980) at Marcoule, where operations were discontinuous. Then, at La Hague, the process was carried out in two stages with two successive pieces of apparatus: an evaporator-calciner and a vitrification furnace (1,100 C). In 1978, the Atelier de Vitrification de Marcoule (AVM) was put into operation industrially to treat the waste from the UP1 plant. The capacity was 40 $L.h^{-1}$ and 15 $kg.h^{-1}$ of glass. At La Hague in 1989 and 1992, the R7 and T7 workshops (for the UP2-800 and UP3 plants, respectively) were launched to treat waste from the UP2-800 and UP3 plants with higher capacities (60 $L.h^{-1}$ and 25 $kg.h^{-1}$ of glass). AVM produced 3,000 containers of 360 kg of glass. The La Hague workshops produced an average of 844 packages of vitrified waste per year and 986 packages of compacted waste per year during the period 2008–2019. In total, as of June 2020, 5,319 CSD-V and U have been shipped to foreign customers and 17,766 CSD-V and U are stored at La Hague in ventilated storerooms, for a total of 23,085 packages.

For the nature of spent fuel casing, France has considered two families of common industrial metals, non- or low-alloyed steels and certain cast irons.

Compared to steel, cast iron offers the possibility of production of the body of the container in a single part, without welds, better resistance to corrosion and a cost that is twice as low. In the end, the option chosen was "spheroidal graphite cast iron". This material is an alloy of iron and carbon with a content of more than 2%, where the crystallization of the carbon has been slowed down to obtain mechanical resistance characteristics that make it a material close to steel and easy to machine. A delicate point was to be mastered: the closing of the container. Three "full thickness" welding techniques were evaluated, TIG welding, electron beam welding and YAG laser welding. The storage container is made up of four cases with stainless steel walls [CEA 08].

5.5.2. *The management of very short-lived radioactive waste*

Most of this waste is hospital waste used as radioactive tracers and has a physical half-life of less than 100 days, managed on site by natural decay. The waste is held for more than 10 times the longest half-life and thus loses more than a thousand times its initial radioactivity. After control, it can be disposed of in conventional channels [AND 18a]. In France, this type of waste is spread over approximately 270 sites and represented a total of 1,736 m^3 of VSLW at the end of 2017. The main radiopharmaceutical companies are Advanced Accelerator Applications, Cyclopharma Laboratories, Cis Bio International and Cis Bio Bioassay [CND 19a]. There are also treatments for thyroid cancers by incorporation of radioactive iodine (brachytherapy). In this case, hospitals collect patients' excreta in storage tanks during the days when patients remain hospitalized.

5.5.3. *Management of very low-level radioactive waste*

France already has a storage site for its very low-level radioactive waste (Morvilliers, Aube). The *Centre industriel de regroupement, d'entreposage et de stockage* (CIRES) was opened in the summer of 2004 and covers 45 hectares in the municipalities of Morvilliers and La Chaise. It receives VLLW mainly coming from the deconstruction of nuclear installations. It is composed of four distinct areas: the storage area, the land disposal area, the storm basin and the industrial area. In the long-term, it should store 650,000 m^3 of waste, mainly from the dismantling of closed French nuclear facilities.

Since 2012, it has been opened up to the consolidation of radioactive waste from non-electronuclear activities and to the storage of some of this waste that does not yet have a final management solution. In 2016, a new sorting and treatment activity dedicated to radioactive waste from non-electronuclear activities was commissioned.

CIRES will operate for about 30 years and will then enter a 30-year monitoring phase.

Very low-level radioactive waste is subject to management requirements and standards. The producer of the waste must obtain exemptions, its waste must be accepted by ANDRA according to the specificities of the CIRES and the producer of the waste must obtain authorization from the DREAL by prefectural decree. The system creates tensions linked to its material consequences (the saturation of storage sites) and to its cost, thus compromising its stability [GAR 14].

The HCTISN was asked by the *Office parlementaire d'évaluation des choix scientifiques et technologiques* (OPECST) to give its opinion on VLLW. The High Committee does not accept the option of implementing unconditional and general release thresholds for all types of VLLW. However, most members of the HCTISN believe that changes are needed in the current management of this waste and, consequently, in the French regulations governing its management. They support the decision of the Ministry of Ecological Transition and Solidarity (MTES) and the ASN of February 21, 2020, which calls for such a change "in order to introduce a new possibility of targeted exemptions allowing, after melting and decontamination, a case-by-case recovery of very low-level metallic radioactive waste". The High Committee also makes several recommendations to optimize public participation and transparency in the context of regulatory change and implementation [HCT 20a, HCT 20b].

The principal sites for the management of VLLW by concentration and confinement are CIRES, then Marcoule, Cadarache, La Hague, the CEA sites of Ile de France, Valduc and Comurhex. Because of the large production of VLLW resulting from the dismantling of nuclear reactors, the CIRES site will be saturated in 2029–2030 [GAR 14].

5.5.4. *Disposal centers for low- and intermediate-level short-lived nuclear waste in France*

Short-lived low- and intermediate-level waste also have storage centers in France at Beaumont-Hague (Manche) and Soulaines (Aube).

The *Centre de stockage de la Manche* (CS), located 20 km northwest of Cherbourg-Octeville in the commune of Digulleville, was in operation for 25 years (1969–1994) and contains 527,214 m^3 of low- and intermediate-level waste. From a simple storage in a trench, the Center quickly evolved. Today, the Manche Center is protected by a watertight cover, the main function of which is to prevent water from percolating onto the packages. It covers 15 hectares. In January 2003, it officially entered the surveillance phase (decree no. 2003-30 of January 10, 2003), with very active surveillance for 10 years.

Since 1992, the *Centre de stockage des déchets de faible et de moyenne activités de l'Aube* (CSFMA) has been storing short-lived low- and intermediate-level waste. It benefits from the 25 years of experience of the CSM. It is located in the communes of Soulaines-Dhuys, Epothémont and Ville-aux-Bois. This waste mainly comes from the operation of nuclear facilities. With a storage capacity of 1 million cubic meters, the CSFMA will be operated for about 60 years and will enter a monitoring phase of about 300 years until the impact of the storage is comparable to that of natural radioactivity (a decrease of a factor of about 1,000 in ^{137}Cs and ^{90}Sr).

Low-level radioactive waste is subject to a management requirement. The producer must be accepted and approved by ANDRA in accordance with the specifications of the CSA and must respect the technical prescriptions contained in the Basic Safety Rules (*règles fondamentales de sécurité*, RFS) issued by the ASN concerning the packages [GAR 14].

5.5.5. *Management of low-level, long-lived nuclear waste in France*

On the other hand, there is currently no storage facility for LL-LLW. LL-LLW represents an intermediate situation because, more than the intensity of its radioactivity, it is the duration that justifies specific storage. There are four types of LLW: radium-bearing, uranium, "irradiated graphite" and historical waste.

5.5.5.1. *Radium-bearing waste*

The first wastes contain radium isotopes. They come from nuclear activities (uranium mines) or from the extraction of rare earths. The radium-containing wastes, which represent 70,000 m^3 of packages to be stored, are currently stored on sites of the *Commissariat à l'énergie atomique* (CEA) in Itteville (Essonne) and at industrial producers such as Rhodia (La Rochelle, Charente Maritime) and Cezus Chimie (Romans, Drôme).

5.5.5.2. *Uranium waste*

Uranium waste or tailings from the conversion of natural uranium (U_3O_8) (RTCU) from Orano at Malvési began from January 1, 2019. These tailings, containing impurities, represent an estimated volume of 55,000 m^3. Uranium waste from before 2019 exists on this site. The ASN considers that long-lived waste should remain in the LL-LLW category [ASN 20b].

5.5.5.3. *Graphite waste*

Graphite waste comes from the nine nuclear reactors of the so-called natural uranium-graphite-gas (UNGG) line that operated between the 1960s and 1990s. These wastes contain carbon-14, which loses half its radioactivity in 5,730 years, and chlorine-36, which has a half-life of 302,000 years. The volume of graphite waste, once conditioned, will represent 100,000 m^3. For its part, the ASN [ASN 20a] estimates this waste classified as LL-LLW at 81,000 m^3.

In the absence of a storage center for long-lived waste, EDF and the CEA have created temporary storage facilities. In 2010, EDF began construction of a temporary facility, called ICEDA, on the Bugey site, which was to condition and store part of the waste from first-generation reactors from the end of 2013 or beginning of 2014. However, the building permit for ICEDA granted on February 22, 2010 by the prefect of the Ain region and the decree authorizing the creation of this facility signed two months later by Prime Minister François Fillon were challenged in court. Work at the construction site was therefore halted for a long time. Recently, on July 29, 2020, the ASN authorized the commissioning of this basic nuclear facility (BNI 173), which is to receive LL-LLW and long-lived intermediate-level metallic waste (LL-ILW).

According to ANDRA [AND 18d], there are five types of graphite waste in France in the category of LL-LLW. These are the graphite liners stored at the EDF site at Saint Laurent-des-Eaux (F5-2-01, 8,424 m^3), the graphite stacks, reflectors and support areas from the former EDF UNGG reactors (F5-2-02, 2,673 m^3), and the graphite liners stored at the CEA at Marcoule (F5-4-01, 1,533 m^3), the graphite waste packages from the CEA's experimental reactors (F5-5-03, eventually by 2030, 1,481 m^3) and the graphite stacks and reflectors from the last military UNGG reactors stored at the CEA at Marcoule (F5-6-01, eventually by 2040, 3,784 m^3), i.e. a total of 17,895 m^3 by 2040 [AND 18a, ZER 18a].

5.5.5.4. *Historic LL-LLW*

Since 2016, the Cires storage facility has been developed to receive waste from activities not related to the production of energy of nuclear origin and not related to the nuclear deterrent force or nuclear propulsion. It essentially responds to the need to take in waste from various historical uses of radioactivity, or from current activities, for which storage outlets are not yet available. It consists, in particular, of objects or products in five main categories: waste from the clean-up of polluted sites (earth or rubble), essentially contaminated with thorium or radium, lightning rod heads (radium-226 and americium-241), sealed or unsealed sources (fire detectors, sources for medical use, etc.), various objects made of depleted metallic uranium and various other wastes (scrap metal, glass, filter cloth, radium, thorium or uranium salts, etc.).

For a transitional period of 30 years, all this waste is stored on site. The storage building, with a surface area of about 2,000 m^2, has been designed to accommodate a volume of 4,500 m^3 of waste packages. The inflow will be 1,000 m^3 at the beginning for the recovery of interim storage, and then routinely 250 m^3 per year.

The decree of February 2010 also sets a maximum radiological capacity for tritium (2 TBq) and carbon-14 (26 TBq) for all the industrial buildings of CIRES, including the storage building. In 2017, the site was under-utilized and did not require a request for extension [AND 17b].

5.5.5.5. *The future storage center for LL-LLW*

Since 2008, ANDRA has been studying the feasibility of creating a storage center for LL-LLW, either by excavation from the surface or in underground tunnels. The solution chosen by ANDRA for this new storage center is to locate it near the ICPE storage center in Aube for VLLW. It would benefit from an incremental approach to study and develop. This will make it possible to adapt to temporal variations in the waste to be stored, both in terms of volume and diversity.

There would thus be several operating campaigns for the center over time, with construction of the storage structures as needed for these successive operating campaigns. Each new campaign would be the subject of an open and progressive governance system with consultations with political leaders, scientists and the public. The process will be carried out in two successive periods, a preliminary programming phase, followed by the launch of the industrial project. The latter was envisaged by ANDRA for the end of 2020 or the beginning of 2021. Since the industrial project will be spread over 12–15 years, the earliest possible date for commissioning the new center would be 2023–2036. The storage of LL-LLW should be spread over a long period, typically about 50 years [AND 17e].

The PNGMDR Orientation Commission issued its opinion at the end of 2020 on the management of this LL-LLW [PNG 20]. It expressed reservations about a possible centralized storage site at Vendeuvre-Soulaines. The next PNGMDR will pursue several objectives that are detailed in six actions.

5.5.6. *Management of long-lived intermediate- and high-level waste in France*

These two categories of waste combine two disadvantages: they are highly radioactive and this radioactivity does not disappear for a thousand centuries. This explains the difficulty of their management.

LL-HLW is the most dangerous waste. Most of it comes from the nuclear power industry. Because of its long life span, it needs to be confined for periods of time of the order of a million years before its radioactivity becomes equivalent to that found in the natural environment. To date, there is no definitive management solution for them. They are currently stored at their respective producers (EDF, CEA, Orano). For Orano, the two main sites are the reprocessing plant at La Hague (Manche) and the Marcoule site (Gard). For the CEA, its storage site is at Cadarache (CEDRA, Conditioning and storage of radioactive waste) which has been operating since 2006. They should be initially stored and then stored in the future deep storage center for this type of waste (Cigéo).

While spent fuel is currently considered in France as radioactive material, in the long-term, it could considerably increase the volumes and activities of LL-HLW radioactive waste. As an example, Table 5.20 gives the characteristics of the very long-lived radionuclides in a discharged spent fuel assembly that has undergone an irradiation of 60 GWj.t^{-1}.

Nature	Radionuclide	Period (year)	Quantity (g.t^{-1} of initial U)	Isotopic content (%)
Minor actinides	^{237}Np	2,140,000	916	100
	^{241}Am	432	490	62.4
	^{243}Am	7,380	294	37.4
	^{245}Cm	8,530	11	8
Fission products	^{79}Se	1,100,000	8	8
	^{95}Zr	1,500,000	1,250	24
	^{99}Tc	210,000	1,410	100
	^{129}I	15,000,000	308	82
	^{135}Cs	2,300,000	769	17
Impurity activation products	^{36}Cl	301,000	2	8
Structure activation products	^{95}Zr	1,500,000	81	0.04
	^{94}Nb	20,300	2	0.4

Table 5.20. *Characteristics of long-lived radionuclides in a spent fuel assembly discharged at 60 GWj.t^{-1} (source: modified from [BON 11])*

5.5.6.1. *History of the search for a definitive solution for LL-HLW in France*

In 1982, the Castaing Commission made recommendations and gave guidelines for the management of irradiated fuels. In 1983, the CEA recommended continental geological disposal as the only reasonable and effective means of isolating waste in the very long-term. Indeed, transmutation can in certain circumstances eliminate transuraniums but not fission products such as ^{99}Tc and ^{129}I. The Commission considers that the criteria for selecting repository sites are stability, permeability, retention, thickness, homogeneity and thermal properties. Four environments are favorable *a priori*: clay, salt, granite and schist[3]. These are, respectively, for France the four departments of Aisne, Ain, Deux-Sèvres and Maine-et-Loire. In 1984, the decision was made to build an underground laboratory. Strong opposition from local populations in several regions forced the political authorities to abandon several projects [CAS 82].

From 1985 to 1987, a new Commission met and the Goguel report was published in 1987. This report lists the criteria for choosing a storage site. The essential criteria are the hydrogeological properties of the site and its geological stability. Other criteria play an important role, in particular the mechanical properties of the host formation and of the formations crossed by the access structures, the geochemical properties of the geological barrier, the respect of a minimum depth, the non-sterilization of the underground resources and the thermal properties of the host formation and of its casing. The group noted that the above criteria do not necessarily lead to the choice of the greatest possible depth. It considers it preferable that the area where the site will be located should offer latitude in the choice of the depth to be retained for the storage. Other recommendations are made concerning the reconnaissance of the site and its follow-up, the work of realizing the storage and the risk of human intrusion [GOG 87]. This report will be discussed and amended. A mini-report will condense the main results [ANO 88].

At that time, the concept of defense in depth for the most dangerous solid waste was adopted with three levels of barriers: the package (matrix, container, biological protection, drum), the filling material (engineered barriers or filler) that immobilizes the water surrounding the package and the geological barrier that ensures hydrogeological isolation, trapping and/or dilution of radionuclides. Similarly, the main phenomena are listed. During operation, three factors come into play: geotechnical stability, prevention of water ingress and thermal effects. Once closure is complete, three phenomena are involved: hydrogeology, the presence of the storage facility and external phenomena. Hydrogeology is involved with underground flows and geochemistry (diffusion, convection, dispersion, retention). The presence of the storage causes a possible inflow of water at the beginning of the storage and an outflow of water in the long-term. The principal external phenomena

3 In the Castaing Commission, the solution of storage in salt was quickly abandoned. The Commission assumed that future generations would be at least as intelligent as our own and that this easy solution for storing waste could attract them. This position upset the High Commissioner of the CEA at the time, Jean Teillac.

that can interfere with the storage site are geological phenomena (climatic, tectonic, and seismic) and human intrusion, which must be prevented for several centuries by choosing a site that is not strategically and economically of interest [ANO 88].

Four geographical areas were explored between 1988 and 1989 (Ain, Aisne, Maine-et-Loire and Deux-Sèvres). Faced with strong opposition from local populations, Michel Rocard, the French Prime Minister, decided on a one-year moratorium and entrusted Christian Bataille (OPECST) with a mission to review the entire system.

The 1991 Bataille law marked a major turning point in French policy on the management of long-lived radioactive waste. In this law, research on the management of radioactive waste is organized along three axes: separation/transmutation, geological disposal and long-term storage. An assessment of the research must be made after 15 years. In addition, this law prohibits the storage of foreign waste and establishes a control of research work carried out by the *Commission Nationale d'Evaluation* (CNE) via the OPECST. Article 13 of this law gives ANDRA the responsibility for designing, establishing and implementing new disposal facilities and, in particular, the creation and operation of underground laboratories for the study of deep geological formations. In the end, only one laboratory will be created, at Bure.

Finally, this law imposes a mediator (decree no. 92-1311 of December 17, 1992) and consultation with local populations. The mediator appointed was Christian Bataille (Order of December 17, 1992, [JOR 92b]). Three sites were preselected, and in December 1998 the government decided to build a laboratory on the Bure site, the Meuse/Haute-Marne underground research laboratory. The mediator, Christian Bataille, submitted his final report to the government of Edouard Balladur in December 1993. The mission was well received by all the actors and led to the proposal of four departments to undertake preliminary geological investigations: Gard, Haute-Marne, Meuse and Vienne [LOC 00]. Of the three sites studied between 1994 and 1997, the Vienne site was the only one with a granitic geology, but the conclusions of the DSIN were unfavorable for this site from the hydrogeological point of view. Similarly, de Marsily [MAR 97] considered that the geological characterization of the Vienne site presented too many prohibitive elements and these justified stopping research on this site. The relevance of installing such an installation not being demonstrated, the installation of an underground laboratory in the Vienne failed in 1998 [PAT 19].

In 1997, the geologist Ghislain de Marsily [MAR 97] estimated that the design of nuclear waste disposal in deep geological formations can only be carried out according to three trends. The first is to give a very significant weight, in the demonstration of safety, to engineered barriers. This is the option taken by Sweden, for example, with its thick copper containers. The second is to look for ways to extract long-lived elements from the waste, to destroy them in reactors or accelerators by transmuting them. The third is to look for a storage site where the containment of non-reprocessed and unseparated waste can be achieved, arbitrarily limiting the scope of safety analyses, in order to be able to carry out a demonstration of safety that is deemed acceptable. This is the case, for example, in Germany, which considers that human intrusion into a salt repository is unrealistic, or in the United States, where the decision to limit the period during which safety will be demonstrated to 10,000 years is tantamount to admitting that the doses that may be delivered by very long-lived radionuclides over the long-term will not be considered. Finally, it should be noted that some countries have simply decided to postpone any decision [MAR 97].

After the selection of the Bure site, the government decided to continue the search for a site in granite and created the *Mission collégiale de concertation granite* with three members (Pierre Boisson, Philippe Huet and Jean Mingasson) by decree on November 19, 1999 [JOR 99]. They gave their report on July 27, 2000. Finally, the law provides for the constitution of a public interest grouping (*groupement d'intérêt public*, GIP) responsible for implementing economic support measures for the establishment of each laboratory. The objective of the Mission collégiale was not to select sites but to conduct a consultation on 15 sites in 16 departments. The difficulties encountered revealed a strong negative reaction from the population in many sites. This refusal was for several reasons. First of all, there was the fear inspired by nuclear energy. Secondly, the fact that storage would be incompatible with the image of quality agricultural products or of a tourist region. Finally, the rejection of the catastrophe which would weigh on the present and future generations. In addition, there was scepticism on the part of certain elected officials as to the advantage, or even the necessity, of carrying out the study of waste storage in two different geological formations. The mediators also encountered a strong mobilization of the anti-nuclear movement. The Granite Mission was a new failure [BOI 00].

The Granite consultation showed the limits of public involvement in a process of experimentation concerning the burial of nuclear waste. Contributing to the "tension" in already highly conflictive social relations, it did, however, have the effect of making the local population aware of the uncertainties surrounding the management of radioactive waste. The scientific debate needed to be socialized because the results of the underground laboratory raised as many technical questions as questions of land use planning. Moreover, this waste management questioned society as a whole about the acceptable risks of any modernization [RUI 04].

The submission of the Argile dossier by ANDRA marked a new stage in 2005. While, following the public debate, the *Commission du Débat Public* noted that the public was voting for permanent storage on the surface or in the subsurface, the 2006 Law chose deep burial as the reference method, while affirming that research would continue in all three areas [PAT 18].

5.5.6.2. *The three lines of research of the 1991 law*

The 1991 law required the search for a solution for long-lived radioactive waste along three lines: separation–transmutation, deep geological burial and temporary storage. In 15 years of studies (1992–2005), the CEA spent 810 million euros (M€) on axis 1, 1,007 M€ on axis 2 and 672 M€ on axis 3 [REV 06].

Axis 1 concerns research on the separation and transmutation of long-lived radionuclides. As early as 2013, the IRSN had serious reservations about whether this method could be an alternative to geological disposal. Despite the progress of ongoing research, it is likely that these operations will not be feasible on an industrial scale in the near future, nor will they be applicable to waste already produced. Moreover, transmutation requires the deployment of a fleet of so-called fourth-generation reactors, the design of which was still being studied in 2013 (since abandoned). In terms of safety, radiation protection and management of radioactive materials and waste, the balance between the gains and constraints brought about by transmutation is very unbalanced, with the gains appearing to be small compared to the strong constraints induced on the fuel cycle. Thus, the IRSN estimated in 2013 that, given the current state of knowledge, transmutation is not in itself a sufficient solution for managing this type of waste [IRS 13e]. Similarly, the ASN [ASN 13], (July 4) considered that the expected gains from the transmutation of minor actinides in terms of safety, radiation protection and waste management do not appear to be decisive, particularly in view of the constraints on fuel cycle facilities, reactors and transport, which would have to use highly radioactive materials at all stages. This would be particularly the case for the transmutation of curium.

Research on deep burial (axis 2) will be developed below (section 5.5.5.3). It has been developed within the Bure underground laboratory, the Cigéo project and its reversibility possibilities.

In France, the temporary storage project (axis 3) was piloted by the CEA. The validity of this type of storage was demonstrated by this organization, which carried out demonstrations with the HERA installation (*Hall d'essais pour l'Entreposage des matières RadioActives*) located at the Marcoule center. This installation includes a subsurface tunnel called Galatée (*GALerie Activités de Tests pour l'EntreposagE*) [SIL 06].

5.5.6.3. *The Bure underground laboratory*

For long-lived radionuclides, which are very harmful substances, the option chosen worldwide is storage in deep geological formations, capable of confining them for several hundred thousand years. In order to validate this solution, ANDRA has dug an underground laboratory 490 m below the town of Bure (between Meuse and Haute-Marne) in a layer of argillite. This laboratory is not the future disposal site but a research facility. In particular, it allows *in situ* observation of 160 million-year-old Callovo-Oxfordian clays.

The underground laboratory consists of surface facilities (administrative offices, workshops, laboratories and a public reception building) covering an area of about 17 hectares and more than 700 m of underground facilities at a depth of 445 and 490 m, located directly in the clay layer. Two wells (main and auxiliary) provide the link between the surface and the underground installations.

The authorization granted to ANDRA to operate an underground laboratory on the territory of the commune of Bure (Meuse) to study the deep geological formations where radioactive waste could be stored was renewed for 18 years in 2011, the first authorization having expired on December 31, 2011 [JOR 11]. The NEA [AEN 01] identifies the role of underground laboratories.

In an underground radioactive waste repository, it is after dissolution in water that atoms or molecules could move according to two mechanisms, "convection" and "diffusion". Diffusion is being tested at Bure using radioactive tracers and monitoring the chemical composition of the pore water (water trapped in the pores of the rock). The results show that the circulation of water is extremely slow in the Bure argillite because of the quasi-impermeability of this rock. Three types of tracers are used: anions (chlorine-36 and iodine-125), cations (sodium-22 and cesium-134) and a neutral radionuclide representative of water (tritium). Three test zones are used at -445 m, -505 m and -540 m, i.e. in the upper, middle and lower zones of the argillite zone, respectively. In addition, the argillite has strong cation retention capabilities (positively charged) within the multiple negatively charged clay sheets. There is a repulsion by the argillite for anions or anionic exclusion [AND 05].

In the underground laboratory at Bure, ANDRA is also experimenting with thermal impact and the effect on desiccation and hydration. It is trying to fill the microcracks resulting from the excavation of the tunnels with dried-out bentonite bricks. In addition, geomechanical studies are conducted with dozens of boreholes equipped with sensors to measure the deformation of the rock around the shafts and tunnels [AND 06b].

5.5.6.4. *The Cigéo project*

The law of June 28, 2006 selects reversible deep geological disposal as the reference solution for the long-term management of long-lived intermediate- and high-level radioactive waste. It entrusts ANDRA with the continuation of studies and research to select a site and design the repository [LAB 11]. At the end of 2009, ANDRA submitted proposals to the government concerning the location and design of the Cigéo industrial geological disposal center.

The surface installations, covering about 300 hectares, include nuclear facilities where waste packages are received, checked and prepared before being placed in storage. The underground storage areas are designed in a modular way to allow the progressive construction of storage cells and the separation of waste according to its characteristics. Their extension after a hundred years of operation will be of the order of 15 km^2. The surface-to-bottom connections would be made by an uncoiled chute for the transfer of packages and by vertical shafts for the work.

ANDRA has defined a zone of interest for in-depth exploration (*zone d'intérêt pour la reconnaissance approfondie,* ZIRA) of 30 km^2. The criteria for selecting this zone are a thickness of the host formation of more than 140 m, a hydraulic head gradient of less than 0.2 mm^{-1}, a depth of the middle of the layer of no more than 600 m, the possibility of installing underground infrastructures perpendicular to the dip of the host layer and a reduced thickness of karstic layers to be crossed in order to establish the connections between daylight and the bottom (shafts or chutes).

According to ANDRA, a site for underground storage of radioactive waste must meet an imperative: the host rock must have very low hydraulic conductivity. This seems to be the case for the Callovo-Oxfordian clay formation at the Bure site. All the studies carried out were necessary to ensure that the water present in and around the site could not, on a very long time scale, transport radioactive elements to the outlets known from previous hydrogeological studies.

Following this submission of proposals by ANDRA, several studies ([IEE 11], IEER, Institute for Energy and Environmental Research; [ANC 12]) have taken up ANDRA's study and note various unknowns. Thus, ANDRA considers that the transport of water in the Bure clay is only diffusive, whereas the IEER considers that it is also convective; this would increase the transported dose by a factor of 10. ANDRA assumes the homogeneity and isotropy of the host layer. However, the stability of the properties of the host layer can be affected by a large number of factors. Some of them cause large-scale disorders, such as seismicity and climate change. Others create local disorders of anthropogenic origin, such as well drilling, tunnel boring, temperature increase (by stored waste) and sealing of the cells. In conclusion, in 2011, for IEER, it seems that ANDRA was too optimistic in its interpretation of the data collected and that there are many uncertainties and gaps in the studies. Their conclusion is that the deadlines for launching the construction of the underground storage center, initially planned for 2015, were unrealistic.

The IAEA is launching a new GEOSAF2 program (Steering Committee chaired by Michael Tichauer, IRSN). This program aims to study the operating methods of long-lived nuclear waste storage centers. The IAEA emphasizes that one of the problems of Cigéo at the beginning of operation was the cohabitation of miners digging the tunnels and nuclear workers placing the packages in the cells. Another problem was the risk of fire in old packages containing bitumen and forming hydrogen.

The 2013 stage

The year 2013 marked an important step for the Cigéo project. A public debate was held from May to October 2013, prior to an inquiry and a request for construction authorization. During this public debate, the IRSN contributed with the drafting of nine fact sheets (inventory of waste intended for Cigéo, reversibility, storage facilities, separation/transmutation, operating phase, package safety, storage sealing, geological barriers and geothermal potential of the site). The position of this organization is summarized in its stakeholder report [IRS 13f]. The IRSN considered that the alternatives to deep geological disposal were not recommendable. Indeed, long-term temporary storage was not appropriate because it knowingly imposed such a burden of control and risk management on future generations, when a permanent solution could be implemented in the near future. In the current state of knowledge, transmutation was not in itself a sufficient solution for managing this type of waste. Moreover, the IRSN considered that it would not provide a convincing gain in safety. Thus, with regard to the alternatives explored following the Bataille law and without prejudice to future major research results, it appeared in 2013 that only geological disposal could constitute an appropriate and sustainable solution for the management of HLW and LL-ILW [IRS 13f].

With regard to the Cigéo project, the IRSN considered that the conditioning of the waste selected was generally appropriate, but had reservations, particularly concerning the conditioning of alpha waste rich in organic matter. Reversibility had positive aspects both at the technical level (monitoring and removal of packages) and at the political level (changes in public concerns). During the operating phase, the studies carried out and the feedback from experience led the IRSN to consider that control of these risks was an achievable objective. However, prevention of the risk of fire remained a major issue that affected the acceptability of bituminous packages in particular. This organization considered that the research carried out by the scientific community confirmed, on the one hand, the favorable characteristics of the Meuse/Haute-Marne site and, on the other hand, that the disturbances generated by the repository could be controllable. However, important additions were still awaited concerning mechanical damage to the host rock or the capacity to seal the structures [IRS 13f].

The 2016 stage

In 2016, ANDRA submitted to ASN three safety option files (DOS) for the Cigéo project for the disposal of high-level and long-lived intermediate-level radioactive waste in deep geological formations concerning the operating phase, retrievability and the post-closure phase [AND 16a; AND 16b; AND 16c]. The ASN asked IRSN to carry out a technical analysis of this file.

For the IRSN, the main issue in examining the DOS was to identify any weak points that could lead to major modifications of the concepts for the construction authorization application (DAC). The IRSN noted that the wastes likely to be stored in Cigéo were generally well identified. It asked ANDRA to pay particular attention to improving the representativeness of the hydrogeological model, to confirming the homogeneity of the host formation and to controlling the damage to the rock and the evolution of the hydromechanical properties of the damaged zones. It also wished to retain certain conservative hypotheses and parameter values more in line with the state of knowledge. In particular, the exclusion of plausible accumulations of events was not relevant. Additional information needed to be provided to ensure that disturbances (bacterial, organic, saline, etc.) were taken into account in a conservative manner in the storage evolution scenarios. It doubted the quality of the low pH concrete in the storage.

The IRSN identified four major points to which ANDRA needed to provide answers in the context of preparing the DAC file. These were the control of risks related to fire in a storage cell for asphalt packages, the consideration of certain accident situations in the operation of the underground installation, the feasibility of monitoring key parameters of Cigéo safety and optimization of the storage architecture from the point of view of safety [IRS 17].

For its part, the ASN [ASN 18a] summarized the various remarks and recommendations, in particular on the bituminous packages, the architecture of the repository, the sizing of the installation to withstand stresses and its monitoring, with a view to its application for creation (DAC).

At the international level, the IAEA gave its opinion on the DOS for Cigéo [AIE 16]. The International Review Team (IRT) positively emphasized the decision to introduce a pilot industrial phase into the process, as well as the preparation of a DOS. It considered that the master operating plan (MOP) was an effective management tool and could play an important role in communication and in

ANDRA's consultations with the ASN, the public and other stakeholders. Its opinion, which was very favorable, offered various suggestions for improvement.

ANDRA proposed its first version of the master plan for operation (PPE). The reference schedule included the inventory of waste taken into account in the Cigéo design studies. The waste intended for Cigéo included ILW and HLW. The reference inventory for Cigéo was 73,600 m^3 of LL-ILW and 10,100 m^3 of HLW.

ANDRA proposed to adopt the following provisional schedule. The start of the pilot industrial phase around 2025, the receipt of the first packages of LL-ILW and HL0 waste around 2030, the transition to routine operation around 2035, and the construction of the surface installations and storage structures for the HL1 and HL2 packages around 2070. Partial closure of Cigéo is planned for around 2070 for the HA0 storage area, 2100 for the LL-ILW storage area and 2145 for the HA1/HA2 storage area. The final closure of Cigéo is planned for 2150. The industrial pilot phase is a period of time in the project that begins with the start-up tests of the facility and ends with the transition to routine operation. Its total duration is estimated at about 10 years, including about four years of inactive testing.

In the context of reversibility, ANDRA emphasizes that the development of Cigéo is incremental, that the operation is flexible, that the installations are adaptable and that recoverability is assured [AND 16e].

The 2020 milestone

In August 2020, ANDRA submitted the public inquiry file (DUP) to the political authorities. It was a large file of about 3,000 pages that included 18 documents (some of which were in several volumes), including a general presentation of the Cigéo project, the characteristics of the repository, as well as legal and administrative documents and documents concerning consultation, the economy, urban planning and the territory. The project impact study was the key document (part 6 in seven volumes) in this dossier. It presented the current state of the environment and the consequences (positive and negative) expected from the project, in terms of the environment (atmosphere, soil, subsoil, water, natural environment, human environment, etc.), health and development of the territory. It also set out all the measures planned by ANDRA to avoid, reduce and compensate for these impacts [AND 20a]. A summary note summarized this DUP [AND 20b]. After the DUP has been obtained, the next step will be to apply for the decree of authorization for creation (DAC).

5.5.6.5. *Reversibility of deep disposal*

The persistent hostility of part of the population to the Cigéo project and the realization that studies on radioactive waste management need to be further developed led Parliament to include in the law, in June 2006, the requirement that the management of high-level radioactive waste be reversible. This reversibility must be possible over a hundred years. It complicates the technical procedures and increases the cost of deep disposal (Figure 5.3), but it will make it possible to benefit from future scientific progress in waste recycling [REA 10]. This hope seems unrealistic, especially for vitrified waste, whose recovery would be very complex, of uncertain safety, and of enormous financial cost.

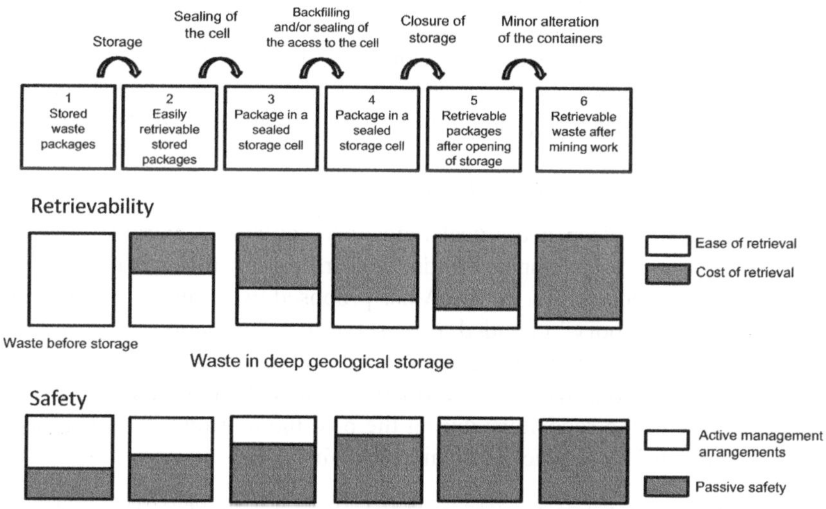

Figure 5.3. *Evolution of the ease of retrieval (recoverability) and passivity (safety) of the Cigéo facility as a function of the various stages of deep geological burial (source: modified from [LAB 11]). For a color version of this figure, see www.iste.co.uk/amiard/radioactive.zip*

5.5.6.6. *Admission of radioactive waste packages to Cigéo*

Packages admitted to a disposal center must meet certain very strict criteria. ANDRA has defined these criteria for the Cigéo project. Following the fire at the American military center of WIPP in 2014, a safety problem appeared with bitumen waste. Indeed, for the latter, the main risk is the possibility of the initiation of internal exothermic reactions. The ASN was aware of this phenomenon and had requested that this waste be taken back [ASN 08]. This type of coating has been used in many countries, even though France is the leading holder of drums of asphalt sludge [CEA 18a]. In France, the total inventory of bituminous waste packages stored on the CEA site at Marcoule represents about 62,000 packages, including 29,000 LL-ILW and 33,000 LL-LLW packages [ASN 19b]. Faced with this problem, the ASN recommends that ANDRA consider a scenario of chain reaction and loss of containment of a package of bituminous waste, after a fire has been extinguished, to monitor the thermal state of the bituminous waste packages and to confirm the absence of the impact of swelling of the bituminous waste packages induced by radiolysis [ASN 19b]. Various industrialists have worked on this problem, in particular the CEA, which has made a synthesis of the state of the art. It notes that there are various thermal treatment processes for bituminous sludge in the world, including several that use a plasma torch (Zwilag in Switzerland, Pluton in Russia and Kozloduy in Bulgaria) [CEA 18a]. The CNE has given a scientific opinion on this subject [CNE 19b]. Since then, the ASN has considered that ANDRA should define, by December 31, 2021, objective criteria for distinguishing bituminous waste that belongs to the LL-LLW system from that belonging to the LL-ILW system, because of its high dose rate and its content of alpha emitters (plutonium-239, plutonium-240, americium-241) [ASN 20b].

5.5.6.7. *Preservation of memory and its intergenerational transmission*

The existence of a deep geological repository containing extremely hazardous high-level radioactive waste for hundreds of years must remain in the memory of many future human generations. ANDRA has proposed two ways of transmitting the memory: “detailed memory” intended to preserve the information necessary for the operator throughout the monitoring phase and “summary memory” intended to provide future generations with the information that will enable them to assess the risks associated with the repository during the post-monitoring phase. But how can “memory documents” be produced in concrete terms?

For the CSM, ANDRA has produced two works of the Center for future generations on permanent paper. A detailed tree-like memoir intended for the management of the Center and for possible future use of the site, in two copies (one at the Center and the other at the *Archives nationales de France*). There is also a summary memorandum in a single volume for the information of decision-makers and the public, suggested by the “Turpin” commission (about a hundred copies kept in various places such as town halls, notary offices, associations, etc.) [AMI 13].

For the eventual storage of long-lived waste in geological formations, ANDRA plans to apply, for the first few centuries, the same type of provisions as for its surface centers. Nevertheless, preserving the memory on a multi-millennial scale requires other types of response. Regarding such periods, what remain of the heritage of the past are rock paintings, parchments and works of art (megaliths, pyramids, ampitheaters, etc.). It was imagined, during international discussions, to build on the surface, above the geological storage centers, markers similar to these works of art. After having resisted erosion or vandalism, will future generations understand the meaning of these structures? Will they be able to identify that these surface structures signal the presence of a geological repository for radioactive waste? [AND 06c, AND 09]. It is also necessary to ensure the sustainability of institutions and to develop vigilance at various levels (local, national and international). The maintenance of the economic and social life of the region also seems to be a solution [REA 10]. International reflections are continuing.

In a program reminiscent of certain science fiction scenarios, linguists, historians, artists and military personnel have spoken. The main result of this program is that there is no efficient way to communicate with a future, mysterious humanity, as temporally distant from us as we are from the first settled villages of Mesopotamia. These future humans will share with us none of language, religion, nor technological and social organization [DEN 16].

5.5.6.8. *The cost of the Cigéo project*

ANDRA has made estimates of the cost of the Cigéo project. The cost of the containers for LL-ILW and LL-HLW is provided in Table 5.21. The annual expenses related to the operation of the center are estimated at 71 million euros (M€) (Table 5.22) and only 3.17 million euros after the center's closure (Table 5.23).

	Summary	
	Number of SCs*	**Total cost (k€)**
LL-ILW	70,609	393,928
LL-HLW	53,959	958,190

Table 5.21. *Summary of the cost of providing storage containers (source: [AND 14b]). *SC: storage container*

Budgets (M€)	**Total gross cost**	**Average annual gross cost**
Workforce	3,609	31.1
Operation	793.5	6.8
Spare parts	784	6.8*
Storage containers and dividers	1,630	14
Utilities	1413.4	12.2
Total	**8,230**	**70.9**

Table 5.22. *Summary of the operating costs of the Cigéo Center (source: [AND 14b]). *Reported over 116 years*

Activities	Cost (M€)
Archives	0.67
Communication	0.20
Monitoring	0.25
Personnel costs	0.75
Taxes	1.30
Total	**3.17**

Table 5.23. *Summary of annual costs and financing after closure of the Cigéo Center (source: [AND 14b])*

5.5.7. *Fierce opposition and the arrival of social problems*

In a public arena, actors are caught up in a game of audience-oriented interaction. They strive to present their cause in order to win as many people as possible to that cause. Opponents of the Cigéo project seek to discredit ANDRA as much as possible by putting their actions into a narrative. For its part, ANDRA has developed a website *Les Arpenteurs* in collaboration with the magazine *Usbeck et Rica* to circulate information that counteracts that of the opponents [CAR 15].

5.5.7.1. *The importance of policy*

Barthe [BAR 09] notes that technologies "are not neutral". This view has even become commonplace in work on scientific expertise, collective risk management, planning conflicts and, more generally, in what is known as the relationship between science and politics. The decisions in these major projects are not simply "technical" choices; they are political, and both are equally important. They are political because they are the result of power struggles, conflicts of interest and negotiations between the different actors who take part in discussions on the subject. But they will also be so because they will concern devices that contain a definition of what deciding means [BAR 09].

Barthe *et al.* [BAR 10] propose a new model of political decision where decisions are revisable and taken at the end of an open debate that allows the groups concerned to intervene in the content of technical choices. This is the notion of reversible political decision.

The French authorities in 1990, in order to break the deadlock on the deep disposal project, modified the time frame, creating a legislative calendar with a research phase independent of the industrial project. They also made an institutional change. But this new political framework remained subordinate to the initial technical framework. In fact, the primacy of the technique was made invisible thanks to a legislative framework [BLA 16].

5.5.7.2. *The public is a major player*

The pursuit of technocratic policies has led in many countries to conflict with affected communities. Since the late 1990s, however, there has been a shift towards more participatory approaches. Authorities have realized the importance of social aspects in radioactive waste management and the need to involve citizens and their representatives in the process. Four countries have taken the participatory turn particularly well: Belgium, Slovenia, Sweden and the United Kingdom [BER 15].

More recently, these mechanisms have been opened up to actors who were previously excluded from the decision-making process, such as representatives of associations, local political life and so-called "ordinary" citizens. Parotte [PAR 18] analyzes the failures and successes of three public management policies, Belgian, Canadian and French, which each led to the choice of burying nuclear waste.

5.5.7.3. *The interests of future generations*

Taking into account the very long-term and the interests of future generations is a concern that is increasingly present in current debates and decisions. In terms of temporality, how should the present generations, who benefit from the production of nuclear electricity, bear the costs of radioactive waste management while taking into account future generations? Traditionally, economic calculations are carried out by means of cost–benefit balances. However, such an approach becomes complex when it comes to projects whose duration and impact horizon may exceed one generation. This is the case for the deep disposal of radioactive waste [DOA 17].

5.5.7.4. *The territory and citizens*

The territory is not a neutral space, but the product of the interaction between space and individuals, who shape it in a material and symbolic sense. It is therefore necessary to reconsider the link to the territory and the effect that it has on the formation of a collective identity [GAR 18].

5.5.8. *A centralized pool as an interim option*

French spent fuel storage capacities are reaching saturation. EDF is therefore required to submit to the ASN the technical and safety options for creating new spent fuel storage capacity. In response to this requirement, in April 2017 EDF requested the ASN's opinion on the safety options for a centralized spent fuel storage pool project. The facility envisaged by EDF is a pool designed to store 10,000 tons of heavy metal (tML), corresponding to about 21,000 fuel assemblies. It would consist of two independent storage pools of identical capacity and design, to be commissioned in stages, with the second pool being built 10 years after the first. At this stage, EDF has not defined a site for the installation. Once the site has been chosen, EDF will have to verify that the levels of hazards chosen for external hazards are appropriate [ASN 19a].

The ASN sought the opinion of its technical partners. On December 20, 2018, the ASN Standing Group of Experts for Laboratories and Plants (*Groupe permanent d'experts pour les laboratoires et les usines*, GPU) issued several recommendations to EDF for the design of this pool (sizing, reinforced concrete lined with a metal skin, monitoring and detection of leaks, control of aging) and to provide an ultimate retention solution in case of leakage (storage basin or transfer channel) [ASN 18c]. For its part, the IRSN [IRS 18d] notes that the design options should lead to a level of safety for this storage pool that is higher than for existing installations. However, the metal liner covering the concrete structure is of delicate design and reduces the safety of the sealing system. Like the GPU, the IRSN regrets the absence of a massive leakage scenario to compensate for such an accident [IRS 18d]. In conclusion, the ASN considers that the general safety objectives and the design options adopted are generally satisfactory [ASN 19a].

According to Laponche [LAP 18], the choice of site for this centralized storage project has already been made and it would be located at Belleville-sur-Loire. More recently, an AFP dispatch (June 30, 2020) highlights the site of La Hague.

5.5.9. *Radioactive waste from the reprocessing of foreign spent fuel*

Commercial recycling of spent fuel is currently carried out only by France and Russia, and in the recent past by the United Kingdom and the United States. In France, reprocessing of foreign fuel began at the UP3-A (1990) and UP2-800 (1994) plants in La Hague (Manche). In total, at the end of 2017, 34,279 metric tons of light water-type spent fuel were processed, of which approximately 69% was for EDF, 16% for German customers, 9% for Japanese customers and the rest mainly for Belgian, Swiss, Dutch and Italian customers. Since mid-2005, Orano has also acquired capacities for processing Research and Test Reactor (RTR) fuel, and contracts have been signed to process spent fuel from French, Australian and Belgian reactors. The entire reprocessing process is a source of large quantities of gaseous (tritium and rare gases) and liquid LLW, which is directly released into the atmosphere and the sea, and of ILW and HLW, in particular fission products, especially strontium-90 and cesium-137, in addition to reusable fissile materials (uranium and plutonium). The recycling of foreign fuel necessarily leads to an increase in the radioactive contamination of the country that carries out the reprocessing. In the United Kingdom, waste substitution is implemented, while in the United States, no return is planned. Since 1977, France's contracts have provided for the return to the client country of part of its radioactive waste. Several laws (December 30, 1991; June 28, 2006) have provided a framework for the treatment of foreign spent fuel and have made the return of foreign waste to its country of origin mandatory. Furthermore, an accounting system for waste (*EXPER* system, *EXPEdition des Résidus*), in particular long-lived waste, allows the waste to be shipped to the client country. Shipments of the first canisters of vitrified waste began in 1995 to Japan and shipments of compacted waste began in 2009 to the Netherlands.

On French territory as of December 31, 2017, the quantity of unprocessed spent fuel present corresponded to 9,970 t of heavy metal, of which 99.6% was French, 0.3% Italian, 0.1% Dutch and less than 0.1% Belgian. For waste packages, on this date, 15,642 CSD-Vs, 400 CSD-Us and 15,608 CSD-Cs were present in the Basic Nuclear Facilities at the La Hague site operated by Orano cycle. Most of these packages were French, except for CSD-C packages, of which 21.9% were German and 9.9% Japanese. For radioactive materials, at the end of 2017, 84 metric tons of uranium (in the form of uranyl nitrate) and 61 metric tons of plutonium (in the form of oxide) were present on the Orano La Hague site, with 15.1% Dutch uranium and 24.3% Japanese plutonium.

For spent fuel that entered France before the law of June 28, 2006, as of December 31, 2017, 5,319 CSD-Vs and CSD-Us were shipped, mostly to Germany, Japan, Switzerland and Belgium, as well as to Australia and the Netherlands. The number of CSD-C packages shipped was 1,180 and approximately 7,060 remained to be shipped to Switzerland, Germany and the Netherlands. For CSD-B packages, 38 have already been shipped and 213 remained to be shipped by the end of 2017.

Spent fuel that entered French territory after June 30, 2006 originated in Belgium, Italy and the Netherlands. Shipment of the packages is scheduled (Table 5.24).

Country	**Spent fuel delivered (tMLi)**	**Number of packages to be shipped**		**Shipping dates**
		CSD-V packages	**CSD-C packages**	
Belgium	0.166	1	0	?-2030
Italy	228.1	79	241	2020–2025
Netherlands	104.8	91 (78)*	90 (66)*	2012–2034

Table 5.24. *Quantity of foreign spent fuel delivered after June 30, 2006, number of waste packages to be shipped and shipment date (source: [ORA 18b]). *Already shipped in 2017*

5.6. The management of specific waste and waste without a channel

A certain amount of radioactive waste does not have a management system for its storage. They are often called "orphan waste" and some of these wastes will become "wards of the Nation" as Zerbib described them in 1983 [ZER 83]. This is currently the case in France with the dismantling of the Bayard plant, which manufactured alarm clocks, with radium-bearing waste [POI 13].

ANDRA [AND 18d] retains only a minimal quantity of nuclear waste without a channel (1,800 m^3 at the end of 2016). This is waste in three categories: containing asbestos, mercury, and organic oils and liquids. They represent only a small part of the total waste (less than 0.1% by volume of the total radioactive waste generated at the end of 2015). They cannot be assigned to existing or planned waste streams because of insufficient characterization, specific physical or chemical characteristics of the waste, or the absence of treatment or conditioning methods for the waste before storage. A two-phase campaign aimed to eliminate these orphan wastes. The first phase (2010–2012) made it possible to identify the wastes qualified as "priority", because of the significant volumes that they represent and the advantages of pooling thoughts and studies between producers, and to define study programs for each of the three categories selected, namely asbestos waste, mercury waste, in the form of metallic mercury or waste containing small quantities of mercury (waste simply soiled with mercury) and certain oils and organic liquids that are not compatible with the acceptance specifications of the Centraco incineration facility. A detailed assessment was drawn up in 2014 and the solutions were included in the 2016–2018 PNGMDR and in the end these three types of waste found their disposal channel.

According to ANDRA, another category poses a problem. This is the activated waste from small producers. The problem associated with activated waste is primarily the difficulty of its radiological characterization. Indeed, among the radioelements generated by the activation process, some, notably pure β emitters, are difficult to measure non-destructively. As non-exhaustive examples, CERN, Ganil, Cisbio or the *Institut Laue Langevin* (ILL, research reactor) are small producers that generate activated parts and regularly submit requests to ANDRA for waste management. Other small producers, such as GIP Arronax or hospitals, are not able to evaluate the activity of activated parts by modeling [AND 18a].

Orphan radioactive waste is more numerous. In fact, we must add to it the historical wastes scattered in a multitude of sites (watchmaking, etc.), tritiated wastes, certain military wastes, wastes from uranium mines and from various plants for the treatment of natural ores, such as the Malvési plant, and submerged waste.

5.6.1. *Management of historical waste*

After the discovery of radioactivity at the end of the 19th century, and particularly during the interwar period, there was a craze for the use of products containing radioactive substances. Radium was perceived as having a destructive power at high doses but a beneficial effect at low doses. The objects concerned were numerous, such as cosmetics, ointments and tablets containing radium, or even devices that enriched water or coffee with radium (radium fountains, radium coffee machines). The most dangerous were the ORUMs (Objects of Radium for Medical Use) with notably needles, tubes, sheaths, "Crowe" probes and radium swabs.

To these must be added radioactive minerals (autunite, pitchblende, monazite) often in private or public collections, and natural salts such as nitrates, acetates and oxides of uranium, uranyl or thorium, and radium nitrates and chlorides that are in many public or private laboratory reserves.

The most numerous radioactive objects come from the clock industry. Indeed, the luminescence of the hands and dials of old models of watches, alarm clocks, compasses, aircraft dials and night vision systems is obtained by the addition of radium-based paint at first, then of tritium [AND 17b].

In France, most of the industries or craftsmen producing these radioactive objects and consequently contaminating their environment with their radioactive waste were in the Ile-de-France and Savoie regions (Annemasse) [ASN 14]. The Swiss watch industry has always been more important than that of France. The Swiss authorities note that the use of tritium in the watch industry to produce the luminosity of luminescent paint can cause significant doses to the installation personnel and leads to the production of radioactive waste that must be disposed of properly. They consider that the use of radioactive luminescent paints is rarely justified [MUR 06].

There are 13 conventional waste storage facilities that have received radioactive waste on a regular or occasional basis. They are located in the municipalities of Angervilliers in Essonne, Argences in Calvados, Bailleau-Armenonville in Eure-et-Loir, Bellegarde in Gard, Champteussé-sur-Baconne in Maine-et-Loire, Le Freney in Savoie, Menneville in Pas-de-Calais, Monteux in Vaucluse, and Le Freney in Savoie, Menneville in Pas-de-Calais, Monteux in Vaucluse, Pontailler-sur-Saône in Côte-d'Or, Saint-Paul-lès-Romans in Drôme, Solérieux in Drôme, Vif in Isère and Villeparisis in Seine-et-Marne.

A dozen other historical waste storage sites are located within or near basic and secret nuclear installations (the A126 highway in Chilly-Mazarin, the Montboucher hillock, the trenches, the internal repository and the concrete basin at Marcoule, the inert waste storage area at Cadarache, the Moronvilliers test pit, the CEN at Saclay, the first six storage sites and the 045 area at Valduc, the Pierrelatte hillock and the Bugey hillock).

In addition to these sites where the volumes of radioactive waste are small, there are four categories of repositories with larger volumes. These are the historical repositories of waste with high natural radioactivity (14 sites), defense storage sites in French Polynesia, dumping sites and uranium mine tailings (17 sites) [AND 18c]. These four categories will be detailed below.

5.6.2. *Storage of tritiated waste*

Because of the extreme mobility of tritium, its storage is tricky. The unfortunate experiment at the Centre de stockage de la Manche (CSM), where part of the tritiated waste escaped from the storage center to contaminate the water table, obliged operators to find special solutions for this radionuclide. Currently, tritiated waste is temporarily stored on various sites, in particular at Valduc (Côte d'Or) and Marcoule (Gard).

The quantity of tritiated waste is relatively large and will become much larger by 2060, especially with the ITER project (Table 5.25) [FRO 11]. However, the majority of tritiated solid waste has no disposal route.

Fromonot *et al.* [FRO 11] distinguish six categories of tritiated waste according to their activity, the fact that degassing is more or less important and the presence or absence of uranium, their more or less strong irradiation and the lifetime of the irradiating radionuclides. This results in specific storage facilities for each of these categories.

Producers	**Storage status (m^3)**	**Inventory (TBq)**	**Cumulative production forecast until 2060**	
			Volume (m^3)	**Inventory (TBq)**
CEA Military applications	3,500	4,200	12,000	1,000
CEA Civil applications	30	2	276	20
Diffuse nuclear	50	220	120	20
ITER	0	0	17,000	33,000

Table 5.25. *Tritiated waste inventory (source: modified from [FRO 11])*

The solution proposed by the CEA is based on decay storage in facilities to be built near the main production sites (Valduc, Marcoule and Cadarache sites) after treatment and conditioning of the waste by the producers. A storage period of about 50 years will be necessary for each of the packages. This period will allow a decrease in activity by a factor of 16 due to natural radioactive decay and will allow ANDRA to create specific centers dedicated to tritiated waste.

Currently, the storage of highly tritiated waste from small producers is handled in the CEA's Valduc site. Eventually, this waste will be extracted from the Valduc center and transferred to the Intermed facility upon opening of this facility, which is currently planned for 2033. This waste from small producers represents a volume of 150 m^3 of solid tritiated inert waste (150 TBq) packaged in 200-liter metal drums. It is mainly pure tritiated waste from radio-luminescent applications [AND 17c].

5.6.3. *Waste of natural origin*

In addition to the mining of uranium and thorium ores, various natural materials are exploited by industries. This is the case, for example, for the extraction of rare earths. Under certain conditions, the tailings from these operations give rise to radioactive waste. A substance becomes a naturally occurring radioactive substance (NORS) when its activity exceeds the exemption thresholds listed in Table 1 of Appendix 13-8 of the French Public Health Code. This threshold is, for example, 10^4 Bq in quantity and 10 $kBq.kg^{-1}$ in concentration for ^{226}Ra.

Currently in France, four hazardous waste storage facilities can receive this waste containing natural radionuclides. These are Villeparisis in the Ile-de-France region (until December 31, 2020), Bellegarde in the Occitanie region (until February 4, 2029), Champteussé-sur-Baconne in the Pays-de-Loire region (until 2049) and Argences in Normandy (until 2023) [ASN 19d].

5.6.3.1. *Mine tailings*

In France, mining deposits were numerous and involved several regions. Tailing storage sites were thus installed near uranium ore processing facilities. These storage sites, ranging from one to several dozen hectares, contain a few thousand to several million tons of tailings (Table 5.26).

A national inventory of former uranium mining sites is updated through the MIMAUSA program (*Mémoire et Impact des Mines d'uranium: Synthèse et Archives*) (https://mimausabdd.irsn.fr/#) making it possible to list and centralize all available information on sites that have been the subject of mining and/or industrial activities.

The sites have been restructured using four main types of processes. An open pit mine is completely or partially filled, with or without underground mining, as at Bellezane and Cellier. An open pit mine with a dike is completely filled as at Brugeaud, Montmassacrot and Lodène. The lowest points of the valley (the thalweg) are connected by a dike, followed by complete filling as at Forez and Bertolière. The waste is placed in a depression surrounded by a merlon or a ring dike with complete filling as at Escarpière, Jouac and Lavaugrasse [PNG 10].

Region	Storage site	Share of total storage (%)	Tailings stored (in thousands of tons)
Languedoc	Le Cellier	12.0	5,967
	Le Bosc (Lodève)	10.0	5,445
Limousin	Bellenaze	3.1	1,646
	Le Bernardan (Jouac)	3.7	1,863
	Brugeaud	25.3	12,547
	Lavaugrasse	15.1	7,488
Pays-de-Loire	L'Ecarpière	22.9	11,350
Rhône-Alpes	Bois-Noirs Limouzal	2.6	1,387
Various regions	Nine other sites	5.3	2,542
Total		**100**	**~50,235**

Table 5.26. *The principal storage sites (more than 1 million tons) for uranium ore processing tailings in France (source: [PNG 17])*

The impact of uranium mines on the environment is far from negligible, because uranium is a metal that is both chemically highly toxic and radiologically very dangerous, even though the risk for humans is controlled. It is therefore essential to know its mobility in the vicinity of mines.

Today, although uranium mining has ceased, France retains traces of its mining history. The mining and processing of uranium ores led to the release of radionuclides into the environment through the discharge of water from the processing of the ore and the evacuation of seepage water from underground structures and open pit mines by means of pumping installations (mine drainage). In addition, these operations have generated two types of mining materials or wastes, waste rock and tailings. Waste rock has two origins. Free waste rock is a material excavated to gain access to the ore, and selectivity waste rock is a material whose uranium content is deemed too low to be mined economically (the cost of a ton of uranium depends on its value on the sterling pound markets). Ore processing tailings are materials remaining after the physico-chemical treatment of the ore to extract the uranium. These materials have uranium contents ranging from a few ppm to 10,000 ppm for tailings from high average grade ores [AND 15]. Moreover, we must not neglect the toxicity of radium-226 and its 14 descendants, several of which are toxic, notably polonium-210, which is very toxic.

The trapping of uranium in the soils and sediments located downstream of former uranium mines is likely to decrease its concentration in surface waters naturally. However, the effectiveness and durability of trapping by the lithosphere depend on the nature and evolution of the uranium-bearing phases and on the physical and chemical conditions of the environment.

Recently, two environments have been studied, the Lac de Saint Clément and hydromorphic soils contaminated by uranium. Lac de Saint Clément is located about 20 km downstream from the former Bois Noirs uranium mine, now used as an underwater storage site for uranium ore processing tailings [STE 18b]. In the sediments of this lake, the progressive reduction of $U^{(VI)}$ to $U^{(IV)}$ beneath the water–sediment interface has been demonstrated. This reduction is strongly controlled by $Fe^{(II)}$ from early diagenesis, in particular through the microbial reduction of structural $Fe^{(III)}$ in clays. For hydromorphic soils rich in organic matter, a stronger reduction of $U^{(VI)}$ is observed, mainly controlled by the water saturation level. Experiments have highlighted the role of organic matter in the control of the mobility of uranium, favoring the remobilization of organic $U^{(IV)}$ complexes under reducing conditions [STE 18b].

Chemical extractions confirmed that approximately 60–80% of U was present as non-crystalline species, highlighting that these species should be considered when evaluating the fate of U in lake environments and the effectiveness of sediment remediation strategies [STE 18d].

The hydromorphic soils (soils saturated with water) studied came from a wetland located near the hamlet of Guern en Tal in the Bubry region (Morbihan, Brittany). The wetland, located about 300 m downstream from the former Ty Gallen uranium mine, is characterized by significant uranium contamination due to the release of mine drainage water during the operation of the mine (up to 4.5 g of U per kg of soil). There is a major redistribution of uranium minerals from U(IV) to U(VI) forming complexes with organic matter. This redistribution of uranium from uranium minerals inherited from the mine's water discharge deposits could result from redox cycling in the vicinity of the water table fluctuation zone [STE 18c].

5.6.3.2. *Natural mineral processing plants*

Some industrial plants process natural ores to extract uranium or rare earths. These plants have the status of classified installations for the protection of the environment (*installation classée pour la protection de l'environnement*, ICPE).

Storage sites for radioactive waste are associated with these plants. In France, this concerns three sites, the RHODIA site at La Rochelle, the CEZUS site at Jarrie and the COMURHEX site at Malvési [PNG 10]. This problem is the subject of discussion in the PNGMDR of 2020.

Rhodia has about 13,700 t of waste in the form of radium-bearing tailings, called RRA (about 1,850 $Bq.g^{-1}$ in alpha and beta activity as of 2002), but only 160 t are at La Rochelle. On the Jarrie site, about 2,000 t of radium-bearing waste are present.

The Malvési site, managed by Orano, has carried out the first stages of conversion of ore into uranium since 1960. Large volumes of historic solid wastes, about 374,000 m^3 of nitrated liquid effluents in the evaporation basins and 70,400 m^3 of sludge, are stored on this site in decantation basins (basins B5 and B6), as well as 282,000 m^3 of mining concentrates, principally in basins B1 and B2 [ASN 19c, ORA 19]. These sludges have a mass activity, mainly in ^{230}Th, of about 200 $Bq.g^{-1}$. Studies are underway to define a definitive management solution for this waste from the RTCU family (*Résidus du traitement de conversion de l'uranium*, Tailings from Uranium Conversion Treatment) in the National Inventory [AND 18a]. Since 2019, new processes have brought the waste into the VLLW and LL-LLW streams.

5.6.4. *Submerged waste*

Ocean dumping has been widely used by some nations such as the United Kingdom, the United States and the USSR (Chapter 3). France participated in only two disposal campaigns for radioactive wastes in the Northeast Atlantic organized by the NEA in 1967 and 1969. In 1967, France dumped 896 metal containers with the waste encased in concrete (0.4 TBq) and 30,700 galvanized steel drums containing thickened liquid effluent treatment sludge (200 TBq). During the 1969 campaign, the dumping took place between 4,000 and 4,600 m and involved 14,800 metal containers containing sludge or concrete (134 TBq).

In addition, the wastes related to the atomic tests in French Polynesia were immersed in three sites, two off the atoll of Mururoa (Novembre and Oscar sites, depth greater than 2,000 m) and one off the atoll of Hao (Hôtel site) [MAR 07]. The Novembre site received, between 1972 and 1975, 76 tons of unconditioned radioactive waste (0.008 TBq). At the Oscar site, 2,580 tons of unconditioned waste were dumped from 1975 to 1982 (0.06 TBq). At these two sites, the radioactivity is due mainly to alpha emitters and especially to plutonium. The Hôtel site received 310 tons of radioactive waste in concrete drums and 222 tons of radioactive waste in bulk. The activity at this site is due to beta–gamma emitters (0.015 TBq) [AND 18e].

Monitoring and control of dumping sites in the Atlantic was carried out until 1977 by each state, then from 1997 to 1995 by the NEA and has ceased since then, with the exception of series of underwater photographs of containers.

The French dumping sites in the Pacific are regularly monitored by the French government with a continuous monitoring of atmospheric aerosols and an annual campaign of environmental samples (Turbo mission). The results of about 200 samples from the various environments of the atolls show only a low level of artificial radioactivity, slightly higher than the natural radioactivity [MIN 15]. In May 2015, an information commission was created at the former nuclear test sites in the Pacific (Mururoa and Fangataufa atolls in French Polynesia) [JOR 15b]. The IRSN also takes measurements in the environment on the Pacific sites.

5.7. French challenges to the radioactive waste management policy

As in most countries, public opposition to radioactive waste management policy is strong and long-standing. In France, local residents' opposition to the construction of underground laboratories has been strong and led to the abandonment of the various projects, with the exception of Bure (see section 5.5.5.1).

The number of associations contesting the use of nuclear energy is significant. Some of them are grouped in collectives such as "*Sortir du nucléaire*".

Certain moral authorities such as the leaders of the Catholic religion are also taking a position. Thus, the bishops of the Bure region have summarized a series of reflections on the deep geological burial project. In the management of waste, there are two areas of responsibility: the safety of the population and the responsibility of the vision of the future. Humanity must show solidarity, and choices and decisions must be guided by the will to serve the "common good". To do justice, it is necessary to deal with these orientations at an international level, and not only at the level of economic decision-makers. The precautionary principle must also be taken into account in the reflection [GUE 12].

Opposition to the closed fuel cycle and the reprocessing of spent fuel are also numerous. The association "Global Chance" is very much at the forefront on this subject.

Contrary to the delirious forecasts of the atomic scientists of the great State bodies, today France consumes less than 500 TWh of electricity, and world nuclear production, far from reaching its predicted peaks, after a peak around the year 2000 of 2,750 TWh, declines each year (2,300 TWh in 2016 and 2,488 TWh in 2017, i.e. 10.3% of the world's electrical production) [CEA 18a]. The uranium shortage has not occurred and its price, supposed to explode as early as 2000, remains low (approximately 49 US$ and only 41 US$ per kilogram in 2017) [GAS 18]. A large part of the spent UOX fuel and all the spent MOX fuel is no longer reprocessed in France because the plutonium and uranium that could have been extracted from it would not have found a buyer either in France or abroad. The result is an unplanned accumulation of these fuels, as well as the reprocessed uranium, which must be stored until the possible emergence of a competitive industrial fast reactor line. The situation is more serious for MOX, which contains much more plutonium than UOX. This has two major consequences: a threefold increase in the risk of a criticality accident and a much smaller number of MOX packages in the pools and their maintenance under water for 150 years, not 50 years [DES 18].

Before 2007, less than half of the spent fuel produced was reprocessed. Since that date, the cumulative quantity of spent fuel in long-term storage has increased by 320 t per year in 2008 and 2009, and then by 150 t per year starting in 2010 [GUI 16b].

Areva respects the return of its waste abroad for LL-HLW. On the other hand, it does not respect this clause for LL-ILW since 5,344 m^3 will remain in France, or 81.1% of the total. Similarly, for SL-ILW, its return is not envisaged, even though the process of reprocessing foreign fuels has generated more than half (53%) of the 120,000 m^3 related to the reprocessing of all PWR fuels received [GUI 16a].

Opposition to the Cigéo project is strong and was expressed, for example, during the last consultation of the PNGMDR in various stakeholder papers.

The associations *Écologie du Carcassonnais, des Corbières et du Littoral Audois* (ECCLA) and *France Nature Environnement Languedoc Roussillon* (Actor notebook n 2) provide an alert on the storage of natural uranium at the former Malvési plant. The site experienced an accident in 2004 involving retention basins whose membranes must be replaced and ASN-prohibited discharges into the Tauran River. The two associations wish to closely follow the progress of the treatment of nitrate waste (TNW), to ask Orano for an improvement in the landscape treatment, and that the mine tailings be considered as waste and find a definitive solution.

ACRO (Actor notebook N 4) considers that the French terminology of "closed fuel cycle" is a deleterious hypocrisy, since in reality no definitive solution is planned or financed for 95% of what comes out of French nuclear reactors, classified as "recoverable" material without being recovered. ACRO points out that the storage of spent fuel is almost saturated (92.6%) and that the recycling rate for this fuel is only 1%. ACRO requests that all "recoverable materials" not recovered after 10 years be classified as radioactive waste and that materials of foreign origin not subject to recovery be returned to their country of origin.

France Nature Environnement (FNE) (Actor notebook No. 5) considers that Cigéo, the deep burial of highly radioactive nuclear waste, is one of the worst solutions. The Association believes that the colossal problem of nuclear waste could influence energy policy and not the other way around. The storage sites for VLLW and LW are saturated, those for LL-LLW and LL-HLW do not exist. The FNE considers that these materials are nuclear waste and that they currently represent 300,000 tons in France. Moreover, the FNE is very sceptical about the adoption by France of release thresholds without downstream control of the materials thus released.

5.8. Conclusions

France's radioactive waste management policy is at a turning point, and decisions will have to be made in the short-term.

5.8.1. *Shortcomings in several categories of radioactive waste*

The quantity of VLLW will considerably increase in the next few years and storage capacities will soon be saturated. A controversy is resurfacing between those who want France to adopt clearance levels for its waste and treat it as ordinary waste and eventually reuse it, and others who want to keep it in the nuclear waste category.

The situation for LLW is more favorable in the short-term. On the other hand, no definitive solution exists in France for the three categories of long-lived waste (LL-LLW, LL-ILW and LL-HLW). The preferred solution for LL-ILW and LL-HLW is deep geological disposal.

It is important to note that several categories of waste do not yet have a management channel in France. These include graphite waste, tritiated waste and radium-bearing waste.

5.8.2. *Recent developments in French nuclear policy*

The 1991 law established three research axes to help find a solution for the management of high-level, long-lived waste. The 2006 law favored axis 2, deep burial.

At the end of 2019, the CEA abandoned the Astrid reactor project, a reactor in the RNR field. Research in this field will be limited to international collaborations with the few countries pursuing projects in this field, such as Japan, Russia and China. One of the consequences of this abandonment is that research on transmutation will be put on hold. This is reinforced by the difficulty and cost of an advanced separation of minor actinides, plutonium and uranium, and a possible solution that can only be envisaged in the long-term. All these reasons have greatly weakened research on separation–transmutation.

The current inadequacy of the French nuclear waste management policy is obvious and a certain contradiction appears with the French energy policy (PPE). One of its contradictions is the use of the large stock of plutonium. Since the abandonment of the fast neutron reactors by France in the medium term, plutonium has no more use. What can be done with it? Should it be reintroduced into a matrix, after having been isolated, to immobilize it and thus make it less toxic for the environment? It is imperative to reflect on this delicate subject.

5.8.3. *Policy change on the closed cycle?*

The hope that France can really close the fuel cycle is diminishing, even in the long-term. Indeed, the need for MOX fuel will decrease in the short-term due to the closure of second-generation PWR reactors, the third generation of reactors, EPRs, are considerably behind schedule, and the abandonment of the RNR option, all of which contribute to France's decision not to reprocess spent fuel, as the majority of nuclear states have done, with the main exceptions of Russia and China.

Currently, the reprocessing of spent nuclear fuel is incomplete and a large tonnage remains in the pools at La Hague. The same is true for plutonium MOX fuel, which cannot be reprocessed on an industrial scale (risk of criticality and high cost). There is thus an accumulation of irradiated fuel in the pools at La Hague and, upstream, in the pools of nuclear power plants in operation [LAP 18].

5.8.4. *Redefinition of radioactive waste and radioactive material*

These various observations lead to the question of the definition of radioactive materials, have they not become radioactive waste?

The ASN, in a decision from October 8, 2020 [ASN 20b], considers it essential that a substantial quantity of depleted uranium be requalified as waste now. The quantity present at the end of 2018 at Bessines-sur-Gartempe (Haute-Vienne) and Tricastin (Drôme) was 318,000 metric tons of heavy metal. Presumably, thorium-bearing substances should also be requalified as waste. For irradiated fuel from the CEA, separated uranium and plutonium from reprocessing, and MOX scraps, the ASN considers that their recoverable character should be reassessed. On the other hand, the ASN considers that the current classification of enriched natural uranium (ENU) spent fuel as material is relevant to the reprocessing policy.

5.8.5. *The cost of waste management*

The average operating costs of storage and warehousing facilities are €137.7M per year. Cumulative investments in these facilities between 2014 and 2017 amounted to €255 million, but could rise to nearly €1.4 billion between 2018 and 2030, and increase by another €1.5 billion between 2030 and 2050. In addition, the deterioration of storage conditions for old waste makes it necessary to take back and condition it for storage, for a total estimated cost of €7.8 billion. Finally, the cost of the Cigéo project (270 km of tunnels buried at a depth of 500 m) was set by decree at €25 billion, but it should be updated regularly and realistically, and estimated for all radioactive waste management scenarios [COU 19a].

In addition, the *Cour des Comptes* has produced a complete summary of the management of French nuclear waste and has made 10 recommendations to the various organizations involved in this field. It asks the DGEC and ANDRA to complete the national inventory by reconciling storage and disposal capacities with current and prospective quantities of materials and waste, to estimate the cost of Cigéo for each of the four scenarios in the national inventory of radioactive materials and waste and to update the costs of the Cigéo reference scenario by taking into account the risks and opportunities of the project in a more realistic way. The *Cour des Comptes* also questions the real industrial perspective of recovery of radioactive materials and the possible storage of spent MOX and ERU [COU 19b].

The contribution of the three French nuclear operators for the construction of Cigéo was approximately 850 million euros for the year 2016 (Table 5.27).

Operators	BNI tax	Special contribution	Additional taxes on waste and storage
EDF	543.6	104.6	112.0
Areva	16.5	6.7	7.2
CEA	6.5	22.8	21.5
ANDRA	5.4		3.3
Other	4.8		1.92
Total	**576.8**	**134.1**	**145.92**

Table 5.27. *Amounts (in millions of euros) of operator contributions for the year 2016 to build the deep geological disposal center (source: [PNG 17])*

6

General Conclusions

6.1. Introduction

The successful management of radioactive waste is essential for the nuclear industry. Indeed, public opinion worldwide is largely unfavorable towards the use of nuclear energy, whether for military or civilian purposes. Moreover, an increasing number of states are abandoning this type of energy following major nuclear accidents. However, this management is not easy and is far from being solved for some categories of radioactive waste. This is particularly true for long-lived waste, as well as for tritiated waste. Solutions must therefore be proposed quickly.

Among the problems to be solved, some should, with good will, find answers in the short-term. This is the case for a unified classification of waste. The creation of safe storage facilities for all wastes can only be solved in the medium term and definitive solutions for long-lived wastes will probably only appear in the long-term. This is why research, both fundamental and applied, must continue intensively. However, it is not ethically responsible to leave it to future generations to manage the waste that was produced to enable the production of electricity that our generation has consumed.

6.2. The main problems concerning radioactive waste

6.2.1. *The problem of multiple classifications*

The fact that each State has its own classification of waste does not help to find common solutions to their management, nor to their inventory. It is urgent to homogenize these various classifications.

6.2.2. *Radioactive waste or nuclear material?*

Depending on whether the nuclear state chooses an open or closed nuclear fuel cycle, the definition of nuclear waste changes significantly. Indeed, for the states advocating the closed cycle, spent nuclear fuel is considered as a fissile nuclear material and not as radioactive waste. This has a major impact on the inventory of waste and on the structure of a future storage center for high-level, long-lived waste.

6.2.3. *Waste without a channel*

In all countries, there is nuclear waste that has no defined storage pathway, nor a constructed center. This is often the case for historical waste such as uranium and thorium mine tailings. It is also the case for waste that is difficult to contain, such as tritiated waste. This type of waste is, for example, in France, the 318,000 tons of depleted uranium, the 31,500 tons of reprocessed uranium and the 8,570 tons of thorium, which have become waste because there is no intended use [AND 20c].

6.2.4. *Long-lived waste*

Among radioactive waste, some is particularly difficult to manage: that is long-lived waste, i.e. waste containing radionuclides with a physical half-life of more than 30 years (31 years for France, excluding strontium-90 and cesium-137 from this category). Indeed, some radionuclides, in particular actinides, have physical half-lives of several hundred years, or even millions of years. As a result, in a hundred centuries, the radioactivity of this waste will still be significant and dangerous for both the environment and humans. Our knowledge does not allow us to predict the fate for the environment, the radioactive packages and the structures of the repository over such long periods of time. Uncertainties remain that must be reduced as much as possible.

6.2.5. *Very low-level waste*

VLLW poses a management problem because of its extremely large volume. This is particularly significant for states that do not accept any release threshold, such as France. This phenomenon is going to be considerably aggravated by the numerous dismantling operations that will take place in the near future. There are two strategies for managing VLLW, either storage on the nuclear production site or centralized storage in a specialized center (such as CIRES in France). To distinguish between these two solutions, Mercat and Lamouroux [MER 18] use the 12 criteria of best available techniques (*meilleure technique disponible*, MTD) retained by France for installations classified for environmental protection (*installations classées pour la protection de l'environnement*, ICPE). The solution of on-site storage seems preferable.

6.3. Innovations in radioactive waste management

Like all research, research in the field of radioactive waste management has priorities that evolve over time. In the past, even though this type of research continued, the main innovations have been related to robotization and the immobilization of radioactive waste by cement, glass or ceramics.

Currently, research is focused on the safety of radioactive transport, separation–transmutation, the aging of the containers and structures of the storage center, the behavior of spent nuclear fuel over time, the possibilities of deep geological burial and its consequences, and communication with the public.

6.3.1. *Research on separation and transmutation*

There are many processes for the separation and conversion of nuclear materials that are aqueous, non-aqueous or pyrochemical. The objective of these processes is to reduce the volume and radiotoxicity of the waste, with varying degrees of success. The purpose of these various processes is the recovery of uranium, the extraction of uranium and plutonium, the co-treatment of actinides, or the extraction of transuranium (TRU). Various options have been studied for extracting various minor actinides (Np, Am, Cm). The separation of other elements such as zirconium (from fuel cladding), cesium and strontium has also been addressed. Most processes use a nitric acid medium, and other media, such as carbonates, are also used. The various processes are at various levels of technical readiness (TRL). More mature processes (higher TRLs) are those based on commercially operated processes, which allow the separation of uranium and plutonium with or without neptunium. Some new processes are in the process of being implemented industrially (TRL 7-8) [OEC 18].

Within the OECD, few states practice reprocessing or have the desire to do so. Reprocessing strategies are very diverse. Thus, reprocessing tends towards a closed fuel cycle (France, Japan). The French strategy envisages multi-recycling of plutonium based on oxides. In Japan, the priority is partitioning and transmutation (P&T). The fuel cycle is classified as open for Korea, the United Kingdom and the United States, which currently operate a once-through fuel cycle. The fuel cycle is classified as open/closed in Russia, where the spent fuel will be reprocessed and recycled to operate several fast reactors [OEC 18].

Reprocessing operations by separating plutonium have possible consequences for the proliferation of nuclear weapons. Therefore, non-proliferation is a key issue that must be taken into account in any development of advanced reprocessing technologies. Different strategies have been explored within the OECD to address this issue [OEC 18].

Only states with a closed fuel cycle, reprocessing spent fuel and practicing advanced radionuclide separation to isolate minor actinides in particular are interested in transmutation. The main research is currently being carried out in China, France, India, Japan and Russia. In Japan, the decommissioning of the Fukushima accident reactors should use transmutation according to the Japanese authorities [NAK 15].

Transmutation necessarily requires a rigorous separation of minor actinides (MA). Concerning the separation of radionuclides, in France, the CEA has been very active and work has been undertaken according to several options. Separation is advanced with successive extractions COEX (U and U-Pu-Np) and SANEX (Am and Cm). Separation can be grouped into GANEX 1 (U) then GANEX 2 (Pu, Np, Am, Cm). Finally, a third option is the recovery of americium only with COEX (U, U-Pu-Np) and EXAm (Am). The EXAm extraction has high yields greater than 99.9% [CEA 15].

In China, a total separation process was developed at Tsinghua University in three extraction cycles: removal of actinides by TRPO (TRalkyl Phosphine Oxide) extraction, removal of ^{90}Sr by dicyclohexyl-18-crown-6 extraction and removal of ^{137}Cs by calixarene extraction (Xu *et al.*, in [IAE 20]).

Recently, it has been demonstrated that transmutation of long-lived actinides in an accelerator-driven system (ADS) is reliable [GOL 18]. However, the solutions, if they materialize, will only be operational in the long-term. Moreover, it should be remembered that if the solutions are conceivable in the long-term, they are completely unrealistic for waste that has already been conditioned, in particular vitrified waste.

6.3.2. *Research on the aging of packaging*

Research concerning waste immobilization is less active, but the possibilities of some nanoparticles fulfill this function. Thus, multi-walled carbon nanotubes (MWCNT) have a high sorption efficiency, a fast rate of sorption, selectivity and reusability, especially for radionuclides of the F block (majority of lanthanides and actinides) [SEN 17].

Over the course of time, whatever it is, a packaging deteriorates in particular by corrosion. It is therefore important to know the rate of degradation, the main factors controlling this aging and to be able to simulate the corrosion numerically or experimentally. The evaluation of groundwater chemistry on the performance of cement-based materials by long-term numerical simulation was proposed by Sampietro *et al.* [SAM 16].

The IAEA [IAE 19b] presented a summary of the main problems concerning the aging of irradiated fuel packages. These problems concern hydrogen embrittlement of the cladding and corrosion of tank welds, with the secondary problems being the degradation mechanisms of concrete overpacks, in particular the freeze–thaw phenomenon and corrosion of reinforcements.

For its part, ANDRA is coordinating research on the treatment and conditioning of two categories of waste: solid technological waste, a mixture of metals and organic materials (vinyls, polyethylene, polymer gloves) and magnesium waste [AND 18b].

The evolution in time of the cements of the packages or the walls of the tunnels raises questions. This evolution is accompanied by an increase in the temperature of the package. An experiment putting cement paste in contact with clayey rock under water-saturated conditions at 70 C and with an increase in the initial temperature led to the dissolution of certain constituents of the package. After one year of interaction, partial decalcification and diffuse carbonation (calcite precipitation) were observed over 800 µm in the cement paste. At the interface, a layer consisting of phillipsite (zeolite), tobermorite (well-crystallized C–S–H), and C–(A)–S–H had formed. Overall, the porosity decreased on both sides of the interface [LAL 16]. Similarly, results on the interaction of cementitious materials with argillite, and their reciprocal influence on their physico-chemical properties have been recently obtained [BER 18].

The presence or absence of bacteria in contact with the steel has a strong influence on any changes in the packages in the deep layers of argillite. An experiment was conducted on this topic. All the techniques showed that the oxic system evolves towards an anoxic system when bacteria are added to the corrosion experiment [MER 16].

Vitrified waste has given much hope in the field of radioactive waste management because the remarkable chemical durability of silicate glass makes it suitable for the immobilization of radionuclides. Their corrosion is slow and the strong decrease in the corrosion rate results from the densification of the weathered layers of the weathered film, leading to the closure of the pores [CAI 08]. The intrinsic dissolution rate of vitrified high-level waste (about 0.02 $g.m^{2}.day^{-1}$) was determined by Fisher *et al.* [FIS 16]. Recently, doubts about the slowness of the corrosion have been raised. Severe localized corrosion has been found at the interfaces between stainless steel and a model nuclear waste glass and between stainless steel and a form of ceramic waste. The accelerated corrosion can be attributed to changes in solution chemistry and local acidity/alkalinity in a confined space, which significantly alter the corrosion of waste form materials and metal cartridges [GUO 20]. The conditions of the experiment are those of the American site containing oxygen, unlike that of Cigéo, and the conclusions cannot be directly extrapolated.

6.3.3. *Research on recycled nuclear fuel and cladding*

After use, spent uranium fuel is considered either as waste for states opting for the open cycle or, on the contrary, as a nuclear material for those practicing the closed cycle of the nuclear fuel cycle. The NEA document [NEA 16b] details the national policy for spent fuel for OECD countries, and the IAEA has organized an international conference where the policy of many countries is outlined [IAE 16b].

MOX fuel currently has heterogeneous microstructures, which increase the release of fission product gases. Therefore, studies of new fabrications are carried out. Thus, the CHROMOX microstructure obtained by doping with Cr_2O_3 presents an improved homogeneity, in particular with smaller primary mixture agglomerates and an increased matrix grain size. Another option (CORAIL-A) consists of putting half of the MOX fuel rods and the rest in the form of UO_2 rods. In contrast, the MIX fuel assembly contains only MOX rods with an enriched uranium matrix that compensates for Pu degradation. However, the most advanced fuel assembly design is GAIA designed by Framatome (Delafoy *et al.*, in [IAE 20]).

The Russians are building a demonstration pilot with a lead-cooled fast reactor (BREST-OD-300) in the Siberian conglomerate of several industries (PRORYV project) that will regroup plants for the manufacturing, recycling, and reprocessing and management of mixed radioactive waste of uranium and plutonium nitrides (Shadrin *et al.*, in [IAE 20]).

The improvement of the resilience of the fuel and of the cladding (zirconium cladding with various coatings and high-density uranium fuel) makes it possible to obtain accident-tolerant fuels (ATF) (Lucas *et al.*, in [IAE 20]).

6.3.4. *Research on deep burial*

Most countries are moving towards deep geological disposal of their high-level and long-lived radioactive waste, and research in this field is active and numerous. Here, only French research will be reported.

6.3.4.1. *Host rock properties*

Argillites are very sensitive to variations in water content, with an increase in their mechanical properties with desaturation and increased suction. Their water retention curve has a classical appearance, with, however, a water damage effect that leads to a stronger swelling after a drying–wetting cycle. The study of the microstructure shows that the clay matrix (50%) can be considered, in the first analysis, as a set of platelets with a well-organized inter-platelet porosity around an average diameter of 32 nanometers (nm), which gives an idea of the average thickness of platelets. A stress release at constant water content results in swelling that results from an internal transfer of inter-platelet free water to the intra-platelet porosity where it is more firmly bound [DEL 16].

During the review of the fifth edition of the PNGMDR (2019–2021), the IRSN prepared 12 short educational sheets to explain its own research on deep geological disposal in argillite. The IRSN's research on nuclear waste storage is divided into two themes: the confining properties of argillite (sheets 1–7) and the disturbances of the confining properties of argillite (sheets 8–12) [IRS 19b].

Diffusion experiments have enabled characterization of the diffusive behavior of different tracers through the clay rock and highlighting of the effects of anionic exclusion. They show that the speed of displacement of the most mobile radionuclides does not exceed a few centimeters per century.

Using natural tracers, the IRSN confirms that diffusion is indeed the dominant transport process within the entire clay layer over periods of several million years. The moment when dilution begins is, however, difficult to determine. As a result, quantitative data are difficult to obtain. The orders of magnitude obtained are nevertheless typical of diffusive processes.

To estimate the transport of radionuclides through fractures, the IRSN uses hydraulic tests. The speed of flow in the fractures is of the order of several kilometers per million years, which is slow but nevertheless much greater than the speed of flow in pristine argillite. The residence time of these waters is estimated to be between 17,000 and 30,000 years (carbon-14 dating). This result is consistent with those obtained by hydraulic tests. Flow over these distances (several kilometers) means that the radionuclides can reach resurgences and there is then a rupture of the containment. It is therefore essential to search for these faults and to find ways of blocking such flows.

The detection of natural discontinuities from the surface can be obtained by two methods: seismic and electrical. Seismic methods do not detect certain so-called "unstripping" faults. It is necessary to complete, if necessary, this approach by a reconnaissance by means of drilling. Electrical methods allow the identification of localized fracture zones in the upper limestones but not in the argillite. However, the fractures present in limestones can sometimes propagate in the argillite; these methods can thus bring qualitative information on the risk of the presence of faults in the argillite.

The detection of natural discontinuities from underground structures is done using seismic methods. The image provided by these methods is consistent with the data acquired in drilling. The seismic methods implemented from the tunnels are therefore a preferred tool for detecting faults within the clay massif.

The sealing of the tunnels will be made with bentonite, which is a clay that swells with water. It is necessary to verify the watertightness and the relationship with the parts of the tunnel. The IRSN began an experiment in 2011. The time required for core swelling was estimated at 4 years. The first modeling tests of the core swelling rate validated the robustness of some models.

Various materials (iron, cement) are used for storage, which can have interactions with the clay rock. Thus, the iron in contact with clay corrodes and in turn changes the physico-chemical properties of the rock. The results show that the steel pellet is corroded to 200 μm thickness after 10 years of interaction. Furthermore, the diffusion of iron in the rock contributed to an increase in its porosity (amount of voids within the argillite). The concrete (pH 13.2) is more alkaline than the argillite (pH 7.5) but has fewer mobile ions (Mg^{2+}, SO_4^{2-}, Cl^- and HCO_3^-). After 20 years, the extension of the disturbance is limited to a few centimeters for the argillite (3–4 cm) and a few millimeters for the concrete.

The excavation of a tunnel results in a significant drop in pore pressure. These results are very useful data to model the mechanical behavior of the rock.

The excavation of tunnels can have an impact in terms of rock displacement in the vicinity of the excavation. Experimentally, the displacements measured in the rock are small (of the order of 0.5 mm). Some models reasonably reproduce the observed measurements.

Variations in humidity and air temperature in tunnels can cause cracks to appear and modify their behavior. Experiments show that the opening–closing of cracks varies from 0 to 3 mm for a temperature varying from 6 to 16°C and a relative humidity varying from 40 to 100%. This is a seasonal phenomenon (closure in summer; opening in winter) [IRS 19b].

6.3.4.2. *The mobility of radionuclides in geological layers*

The mobility of radionuclides in a rock depends on the permeability and the presence of faults or cracks.

The tests also show that the permeability to water of the site is about 10^{-20} to 10^{-21} m² and much lower than the permeability to gas. Finally, we note that the Biot[1] coefficient is not significantly affected by thermal effects, but that these effects play on the relative permeability to gas of the clay [YUA 17].

Deep clay rocks with very low permeability (10–20 m^2), such as the Callovo-Oxfordian (COx) Clay in France or the Opalinus Clay in Switzerland, are potential host rocks for geological disposal of radioactive waste. Pore-elastic calculations with the boundary conditions of the improved drainage system determine the order of magnitude of the axial deformation rates, allowing a good drainage during triaxial shear (6.6×10^{-8} s^{-1}). All these results reduce the uncertainty concerning the thermo-hydromechanical (THM) properties of the clays and should allow a better estimation of the near-field response of the galleries during the different stages of their service period [BEL 17].

A model using a phenomenological law and linking the intrinsic permeability of the material to the internal variables of the mechanics was used by ANDRA. This model was applied to the underground structures at the Bure site in order to understand the alteration mechanisms of the hydromechanical properties of the Callovo-Oxfordian Clay (COx) around the tunnels and storage cells caused not only by the excavation operations but also by the overpressures due to the production of hydrogen gas following the corrosion of the metal parts of waste modules [MAH 17].

Opalinus clay containing kaolinite (OPA) is the proposed host rock for radioactive waste disposal in Switzerland. However, the presence of tectonic faults intersecting the OPA formation has called into question the long-term safety performance of the underground repository due to the possibility of earthquakes caused by fault instability. The study by Orellana *et al.* [ORE 18] confirms the seismic safety of the OPA formation for a nuclear waste repository.

---

1 The Biot coefficient takes into account many elastic and acoustic parameters of the porous material.

Excavation of subsurface excavations usually results in the creation of an excavation disturbed zone (EdZ) or excavation damaged zone (EDZ) as defined by Tsang *et al.* [TSA 05], resulting from micro- and macro-fracturing induced by the redistribution and rearrangement of initial stresses [SOU 17].

For his part, Carlioz [CAR 17] studied the birth of cracks during the excavation of a tunnel in a clay material and the nucleation of a crack of macroscopic dimension. He found that the nature of the cracks likely to nucleate was closed.

6.3.4.3. *The presence of bacteria in clays*

The microbial community present in the host rock or buffer material may compromise the effective performance and safety of the waste disposal system. In the Czech Republic, bentonite from the locality of Černý vrch should be used as a buffer material. Both samples (bentonite and host rock) collected were inhabited by relatively similar bacteria. Beta- and alpha-proteobacteria dominated both samples. In addition, chemolithotrophs, including *Thiobacillus*, *Gallionella* and *Nitrosomonas*, capable of oxidizing NH_3, Mn^{2+}, Fe^{2+} and S^{2-} were also present [SHR 16].

Within the NEA, the Salt Club has supported and overseen the characterization of rock salt as a potential host rock for deep geological deposits. It has conducted a study of the microbiology of these formations and extended it to other host rocks (granite, basalt, tuff and clay) in Europe and the United States. The study found that some uncertainty remains about the effects of microorganisms on the performance of salt-based storage [NEA 18b].

In Russia, the study of subsurface microorganisms at FSUE MCC (Mining and Chemical Combine) and JSC SCC (Siberian Chemical Combine) liquid radioactive waste (RW) deep storage facilities was conducted in 1998–2016. Injection of liquid radioactive effluent into aquifers increased the population of microorganisms, as well as the rates of anaerobic microbial processes resulting from the use of organic and inorganic components present in the waste [SAF 18].

6.3.4.4. *Transmission of information to future generations*

Several international standards apply to the signaling of nuclear danger, the famous black and yellow “clover” (ISO 361) supplemented since 2007 by a second symbol (ISO 21482) in order to be understood in a more universal way. But for high-level, long-lived waste, the solution that is mostly accepted today, which is deep geological disposal, raises a question of communication. Indeed, if the waste is confined underground for millions of years, how can we warn of the existence and location of these sites in order to avoid any future human intrusion, whether voluntary or accidental? [ROC 18]. The message intended for future generations can take three options: symbolic, indexical or iconic (see Chapter 4).

6.3.5. *Communication to the public*

It has become apparent that the support of local populations for the creation of a radioactive waste storage center is fundamental. Many projects have had to be abandoned after fierce public opposition. Although nuclear energy is recognized as a social problem, and thus concerns all humans, there is still insufficient public debate on the problems posed by this form of energy. This is true in Japan [YOS 15], but also in many other countries.

Table 6.1 summarizes the various options available to managers associated with their functions and limitations.

In conclusion, radioactive waste management is not an easy step. State policies are very diverse. Most states have created specific institutions to manage radioactive waste. This is its sole purpose, such as ANDRA in France, or it combines this function with that of coordinating decommissioning and dismantling, such as the Nuclear Decommissioning Authority (NDA) in the United Kingdom. However, no State currently has a complete policy for the management of its radioactive waste. The shortcomings are particularly glaring for high-level, long-lived waste and for spent nuclear fuel. This is also the case for historic waste, such as military waste and waste from the early stages of the nuclear fuel cycle, which is often poorly conditioned and without final disposal.

In the management of radioactive waste, one of the problems that the authorities must resolve is the strong social rejection by the local population with regard to the storage of this category of waste. This aspect of the situation is as important as the technical difficulties, regardless of the geological qualities of the site chosen.

Another difficulty stems from the significant cost of this management. Here again, the solutions vary considerably from one country to another. But the main obstacle to management is the fact that radioactive waste management is a process that takes place over a long period of time at all stages, from the design of the repository to its construction, operation and monitoring after its closure. Design and construction can be counted in decades, operation in a century and monitoring in several centuries. However, the projection into the future is full of uncertainties, the main ones being the integrity of the artificial and natural barriers if they will still be sufficient after 10,000 years of existence.

Other questions remain, such as whether transmutation will be able to pass the industrial stage or, for states that have chosen the closed fuel cycle, whether reprocessing of spent fuel is necessary.

Options	Functions	Limitations
Near-surface storage facility without EBS	Excavated trenches covered with a layer of soil. Simple and inexpensive	Suitable for VLLW and LLW only. Erosion, intrusion and percolation of rainwater may affect performance
Near-surface storage facility with EBS	Multi-barrier approach to improve disposal safety	Suitable for LLW. Long operating experience. Limited amount of long-lived radionuclides. Erosion, intrusion and rainwater percolation can affect performance
Drilling and intermediate depth cavities	The depth is adequate to eliminate the risk of erosion, intrusion and percolation of rainwater. Possibility of using existing disused cavities and mines. Simple and inexpensive (drilling)	Geological barriers are site dependent
Geological storage facility with deep boreholes	Suitable for all categories of waste. Improved containment	Site-dependent geological formations. High cost. Complex technology involved. In-depth safety and performance analyses

Table 6.1. *Functions and limitations of some radioactive waste storage options (source: [OJO 14]). EBS: engineered barrier system*

List of Acronyms

ACRO	*Association pour le contrôle de la radioactivité dans l'Ouest* (French Association for the Control of Radioactivity in the West)
ADS	Accelerator-Driven System
AE	*Autorité environnementale* (French Environmental Authority)
AECL	Atomic Energy Canada Limited
AGR	Advanced Gas-cooled Reactor
ANCCLI	*Association national des comités et des commissions locales d'information* (French National Association of Local Information Committees and Commissions)
ANDRA	*Agence national pour la gestion des déchets radioactifs* (French National Agency for Radioactive Waste Management)
ASN	*Autorité de sûreté nucléaire* (French Nuclear Safety Authority)
BNI	Basic Nuclear Installation
BWR	Boiling Water Reactor
CANDU	CANada Deuterium Uranium
CCEA	*Commission de contrôle de l'énergie atomique* (Atomic Energy Control Board)

CEA	*Commissariat à l'énergie atomique et aux énergies alternatives* (Atomic Energy and Alternative Energies Commission)
Cigéo	*Centre industriel de stockage géologique* (Industrial Geological Storage Center)
CIRES	*Centre industriel de regroupement, d'entreposage et de stockage* (Industrial Center for Regrouping, Storage and Warehousing)
CLAB	Central Interim Storage Facility for Spent Nuclear Fuel
CLI	*Comité local d'information* (Local Information Committee)
CNDP	*Commission nationale du débat public* (National Commission for Public Debate)
CNE2	*Commission nationale d'évaluation* (National Evaluation Commission)
CNSC	Canadian Nuclear Safety Commission
COVRA	*Centrale Organisatie Voor Radioactief Afval* (Central Organization for Radioactive Waste)
CSD-C	*Colis standard de déchets compactés* (Standard Compacted Waste Package)
CSD-V	*Colis standard de déchets vitrifiés* (Standard Vitrified Waste Package)
CSM	*Centre de stockage de la Manche* (Manche Storage Center)
DAC	*Demande d'autorisation de construction* (Request for Authorization to Build)
DAM	*Direction des applications militaires* (Directorate of Military Applications)
DFGP	*Dépôt dans des formations géologiques profondes* (Deep Geological Repository)
DOE	Department of Energy

DOS	*Dossier d'options de sûreté* (Safety Options File)
DSC	Dry Shielded Canister
DSIN	*Direction de la sûreté des installations nucléaires* (Directorate for the Safety of Nuclear Installations)
DSND	*Délégué à la sûreté nucléaire et à la radioprotection pour les activités intéressant la Défense* (Delegate for Nuclear Safety and Radiation Protection for Defense-related Activities)
EBS	Engineered Barrier System
EDF	*Électricité de France*
ENU	Enriched Natural Uranium
EOS	Extended Optimized Storage
ERU	Enriched Reprocessed Uranium
EU	European Union
EW	Exempt Waste
FSUE	Federal State Unitary Enterprise
GCR	Gas-Cooled Reactor
HCTISN	*Haut comité pour la transparence et l'information sur la sécurité nucléaire* (High Committee for Transparency and Information on Nuclear Safety)
HLW	High-Level Waste
HWR	Heavy Water Reactor
IAEA	International Atomic Energy Agency
ICEDA	*Installation de conditionnement et d'entreposage de déchets activés* (Installation for the Conditioning and Storage of Activated Waste)

ICPE	*Installation classée pour la protection de l'environnement* (Installation Classified for the Protection of the Environment)
IEER	Institute for Energy and Environmental Research
ILRW	Intermediate-Level Radioactive Waste
ILW	Intermediate-Level Waste
IRSN	*Institut de radioprotection et de sûreté nucléaire* (French Institute for Radiation Protection and Nuclear Safety)
ISFSF	Interim Storage Facility for Spent Fuel
L/ILW	Low/Intermediate-Level Waste
LDC	London Dumping Convention
LLRW	Low-Level Radioactive Waste
LLW	Low-Level Waste
LTS	Long-Term Storage
MACSTOR	Modular Air-Cooled STORage
Magnox	MAGnesium Non-OXidizing
MOX	Mixed Oxide Fuel
NDA	Nuclear Decommissioning Authority
NEA	Nuclear Energy Agency
NFC	Nuclear Fuel Cycle
NRC	Nuclear Regulatory Commission
NWMO	Nuclear Waste Management Organization
OECD	Organization for Economic Cooperation and Development

ONDRAF	*Organisme national des déchets radioactifs et des matières fissiles enrichies* (Belgian National Agency for Radioactive Waste and Enriched Fissile Materials)
OPECST	*Office parlementaire d'évaluation des choix scientifiques et technologiques* (French Parliamentary Office for the Evaluation of Scientific and Technological Options)
OPG	Ontario Power Generation
PNGMDR	*Plan national de gestion des matières et déchets radioactifs* (French National Plan for the Management of Radioactive Materials and Waste)
PWR	Pressurized Water Reactor
RBMK	*Reaktor Bolshoy Moshchnosti Kanalnyi*
RK&M	Preservation of records, Knowledge and Memory
RWM	Radioactive Waste Management
RWMC	Radioactive Waste Management Committee
SBNI	Secret Basic Nuclear Installation
SFR	Repository for Short-Lived Radioactive Waste
SKB	*Svensk Kärnbränslehantering AB*
SLW	Short-Lived Waste
SNF	Spent Nuclear Fuel
STUK	Finnish Radiation and Nuclear Safety Authority
tHM	tons of Heavy Metal
UCTR	Uranium Conversion Treatment Residues
UNGG	*Uranium naturel graphite gaz* (Natural Uranium-Graphite-Gas)
UOX	Uranium OXide-based fuel

URL	Underground Research Laboratories
USAEC	United States Atomic Energy Commission
VLLW	Very Low-Level Waste
VSL	Very Short Life
VSLW	Very Short-Lived Waste
WAC	Waste Acceptance Criteria
WIPP	Waste Isolation Pilot Plant

References

[AAK 96] AAKROG A., "Inventory of nuclear releases in the World", in LUYKX F.F., FRISSEL M.J. (eds), *Radioecology and the Restoration of Radioactive-contaminated Sites*, Kluwer Academic Press, Dordrecht, 1996.

[AAK 97] AAKROG A., DAHLGAARD H., NIELSEN S.P. *et al.*, "Radioactive inventories from the Kyshtym and Karachay accidents: Estimates based on soil samples collected in the South Urals (1990–1995)", *The Science of the Total Environment*, vol. 201, pp. 137–154, 1997.

[AAK 03] AAKROG A., "Input of anthropogenic radionuclides into the World Ocean", *Deep-Sea Research*, vol. 2, no. 50, pp. 2597–2606, 2003.

[AAL 16] AALTONEN K., KUJALA J., HAVELA L. *et al.*, "Stakeholder dynamics during the project front-end: The case of nuclear waste repository projects", *Project Management Journal*, vol. 46, no. 6, pp. 15–41, December 2015/January 2016.

[ACR 12] ACRO, Le contrôle des rejets hospitaliers, Report, ACRO, p. 9, 2012.

[ADE 20] ADEME, "Déchets chiffres-clés", *Faits et chiffres*, 2020.

[AEN 00] AEN, Évacuation géologique des déchets radioactifs. Bilan des dix dernières années, Report, AEN, 2000.

[AEN 01] AEN, La gestion des déchets radioactifs : le rôle des laboratoires souterrains, Report, AEN, 2001.

[AEN 02a] AEN, La réversibilité et la récupérabilité dans la gestion des déchets radioactifs, Une réflexion à l'échelle internationale, Report, AEN, 2002.

[AEN 02b] AEN, *Réaménagement de l'environnement des sites de production d'uranium*, OECD, AEN, Paris, 2002.

[AEN 03] AEN, "Programme français de R-D sur le stockage géologique de déchets radioactifs", *Revue internationale par des pairs du Dossier 2001 Argile*, 2003.

[AEN 04a] AEN, Dossier de sûreté post-fermeture d'un dépôt en formation géologique, Nature et finalité, Report, AEN, 2004.

[AEN 04b] AEN, "Gestion des échelles de temps dans l'évaluation de la sûreté en phase post-fermeture, Enseignements tirés de l'atelier d'avril 2002 à Paris, France", *NEA*, no. 5148, 2004.

[AEN 05] AEN, "La prise de décision par étapes dans la gestion à long terme des déchets radioactifs. Expérience, résultats et principes directeurs", *NEA*, no. 6039, 2005.

[AEN 06a] AEN, "Comprendre les attentes de la société dans la gestion des déchets radioactifs et s'y adapter. Enseignements principaux et expériences du Forum sur la confiance des parties prenantes", *NEA*, no. 5297, 2006.

[AEN 06b] AEN, Sûreté du stockage géologique de déchets radioactifs HAVL en France, International examination from "Dossier 2005 Argile" peers about Callovo-Oxfordian formation, *NEA*, no. 6179, 2006.

[AEN 07a] AEN, "Créer un lien durable entre une installation de gestion de déchets et sa collectivité d'accueil. Valeur ajoutée à travers la conception et les processus", *NEA*, no. 6177, 2007.

[AEN 07b] AEN, "Changements culturels et organisationnels dans les organismes de gestion des déchets radioactifs. Enseignements tirés", *NEA*, no. 6181, 2007.

[AEN 10a] AEN, "Partenariats pour la gestion à long terme des déchets radioactifs. Evolution et pratique actuelle de treize pays", *NEA*, no. 6824, 2010.

[AEN 10b] AEN, "Rôle des informations géoscientifiques dans le dossier de sûreté pour la gestion des déchets radioactifs. Principales conclusions du projet AMIGO", *NEA*, no. 6396, 2010.

[AEN 10c] AEN, "Réglementation et lignes directrices pour l'évacuation des déchets radioactifs en formation géologique. Revue de la littérature et des initiatives de la dernière décennie", *NEA*, no. 6406, 2010.

[AEN 12a] AEN, "Stockage géologique des déchets radioactifs : engagement national, participation locale et régionale", Collective statement of the Radioactive Waste Management Committee of the OECD Nuclear Energy Agency, Adopted in March 2012, *NEA*, no. 7037, 2012.

[AEN 12b] AEN, "Réversibilité des décisions et récupérabilité des déchets radioactifs. Éléments de réflexion pour les programmes nationaux de stockage géologique", *NEA*, no. 7105, 2012.

[AER 06] AERB, Near surface disposal of radioactive solid waste, Guide no. AERB/NRF/SG/RW-4, AERB Safety Guide, Government of India, May 2006.

[AER 07a] AERB (Atomic Energy Regulatory Board), Management of radioactive waste, Code no. AERB/NRF/SC/RW, AERB Safety Code, Government of India, August 2007.

[AER 07b] AERB (Atomic Energy Regulatory Board), Management of radioactive waste from mining and milling of uranium and thorium, Guide no. AERB/NF/SG/RW-5, AERB Safety Guide, Government of India, August 2007.

[AHN 15] AHN J., NAKAYAMA S., "Performance assessment of the radioactive waste disposal system", in NAGASAKI S., NAKAYAMA S. (eds), *Radioactive Waste Engineering and Management*, Springer, Tokyo, 2015.

[AIB 14] AIB (Accident Investigation Board), Accident investigation report, Underground salt haul truck fire at the waste isolation pilot plant, February 5, 2014, Report, U.S. Department of Energy, Office of Environmental Management, March 2014.

[AIB 15] AIB (Accident Investigation Board), Phase 2, radiological release event at the waste isolation pilot plant, February 14, 2014, Report, U.S. Department of Energy, Office of Environmental Management, April 2015.

[AIE 16] AIEA (*Agence internationale de l'énergie atomique*), Revue internationale par les pairs sur le "dossier d'options de sûreté" du projet de stockage de déchets radioactifs en couche géologique profonde : Cigéo, Peer review report, AIEA, Paris, France, November 2016.

[AIE 17] AIEA (*Agence internationale de l'énergie atomique*), "Convention commune sur la sûreté de la gestion du combustible usé et sur la sûreté de la gestion des déchets radioactifs. Aperçu général", *Brochure IAEA*, IAEA, Vienna, 2017.

[ALD 20] ALDEN A., "Why not dispose of waste in ocean trenches?" *ThoughtCo*, available at: www.thoughtco.com/dont-dispose-waste-in-ocean-trenches-1441116, February 11, 2020.

[AMI 13] AMIARD J.C., *Le risque radioactif. Devenir des radionucléides dans l'environnement et impacts sur la santé*, Lavoisier, Tec&Doc, Paris, 2013.

[AMI 18] AMIARD J.C., *Military Nuclear Accidents: Environmental, Ecological, Health and Socio-economic Consequences*, ISTE Ltd, London and John Wiley & Sons, New York, 2018.

[AMI 19] AMIARD J.C., *Industrial and Medical Nuclear Accidents: Environmental, Ecological, Health and Socio-economic Consequences*, ISTE Ltd, London and John Wiley & Sons, New York, 2019.

[AMI 21] AMIARD J.C., *Disarmament and Decommissioning in the Nuclear Domain*, ISTE Ltd, London and John Wiley & Sons, New York, 2021.

[ANC 12] ANCCLI, Analyse du rapport final préparé par l'IEER pour le CLIS de Bure. Examen critique du programme de recherche de l'ANDRA pour déterminer l'aptitude du site de Bure au confinement géologique des déchets à haute activité et à longue vie, Report 08_02_12, ANCCLI's Scientific Committee, 2012.

[AND 05] ANDRA, "La diffusion des radioéléments dans la roche", *La vie du Labo*, no. 28, pp. 5–7, 2005.

[AND 06a] ANDRA, "Recherches en profondeur. Où en sont les expérimentations ?", *La vie du Labo*, no. 31, pp. 5–7, Springs 2006.

[AND 06b] ANDRA, "Nouvelle loi sur les déchets radioactifs. Les conséquences pour le laboratoire souterrain de Meuse/Haute Marne", *La vie du Labo*, no. 32, pp. 5–6, Summer 2006.

[AND 06c] ANDRA, Préserver la mémoire des stockages pour les générations futures, Report, ANDRA, 2006.

[AND 09] ANDRA, Stockage réversible profond, étape 2009. Options de réversibilité du stockage en formation géologique profonde, Report, ANDRA, 2009.

[AND 14a] ANDRA, "Evaluation des coûts afférents à la mise en œuvre des solutions de gestion à long terme des déchets radioactifs de haute et de moyenne activité à vie longue", *Proposition de l'ANDRA*, ANDRA.398.B, vol. 1, October 2014.

[AND 14b] ANDRA, "Evaluation des coûts afférents a la mise en œuvre des solutions de gestion a long terme des déchets radioactifs de haute et de moyenne activité à vie longue", *Proposition de l'ANDRA*, ANDRA.398.B, vol. 2, October 2014.

[AND 15] ANDRA, Inventaire national des matières et déchets radioactifs, Summary report, ANDRA, 2015.

[AND 16a] ANDRA, Dossier d'options de sûreté partie après fermeture (DOS-AF), Technical document, ANDRA, CG-TE-D-NTE-AMOA-SR2-0000-15-0062/A, 2016.

[AND 16b] ANDRA, Dossier d'options de sûreté partie exploitation (DOS-EXPL), Technical document, ANDRA, CG-TE-D-NTE-AMOA-SR1-0000-15-0060/A, 2016.

[AND 16c] ANDRA, Dossier d'options techniques de récupérabilité (DORec), Technical document, ANDRA, CG-TE-D-NTE-AMOA-RV0-0000-15-0059/A, 2016.

[AND 16d] ANDRA, Inventaire national des matières et déchets radioactif. Bilan à fin 2015, Report, ANDRA, 2016.

[AND 16e] ANDRA, Proposition de plan directeur pour l'exploitation de Cigéo, Report, ANDRA, Version 1, April 2016.

[AND 17a] ANDRA, Tout savoir sur la gestion des déchets radioactifs, Report, ANDRA, 2017.

[AND 17b] ANDRA, Connaître et reconnaître les objets radioactifs, available at: https://www.andra.fr/sites/default/files/2019-02/plaquette_objets_radioac_exe_nov2017_BD_newblocadresse.pdf, November 2017.

[AND 17c] ANDRA, PNGMDR 2016-2018 stratégie envisagée par l'ANDRA pour la gestion des déchets tritiés solides des petits producteurs dans l'attente de la mise en service des installations d'entreposage prévues pour ITER, Technical document, ANDRA, DG/17.0253, 2017.

[AND 17d] ANDRA, "PNGMDR 2016-2018. Analyse des capacités d'entreposage du centre industriel de regroupement, d'entreposage et de stockage, Technical document, ANDRA, DG/17.0240, 2017.

[AND 17e] ANDRA, PNGMDR 2016-2018. Date prévisionnelle de mise en service d'un nouveau centre de stockage dans l'Aube, Technical document, ANDRA, DG/17.0163, 2017.

[AND 18a] ANDRA, PNGMDR 2016-2018. Etat d'avancement des études relatives au traitement des déchets actuellement sans filière, Technical document, ANDRA, DG/17.0254, 2018.

[AND 18b] ANDRA, Traitement et conditionnement des déchets radioactifs, Report, ANDRA, Dossier 02, 2018.

[AND 18c] ANDRA, Inventaire national des matières et déchets radioactifs, Summary report, ANDRA, 2018.

[AND 18d] ANDRA, Catalogue descriptif des familles, Report, ANDRA, p. 248, July 6, 2018.

[AND 18e] ANDRA, Les déchets radioactifs immergés, Thematic dossier, ANDRA, 2018.

[AND 19] ANDRA, Inventaire national des matières et déchets radioactifs. Les essentiels, Report, ANDRA, 2019.

[AND 20a] ANDRA, Dossier d'enquête publique préalable à la déclaration d'utilité publique du centre de stockage Cigéo, Report, ANDRA, July 2020.

[AND 20b] ANDRA, La demande de déclaration d'utilité publique du projet de centre de stockage Cigéo, Summary note, ANDRA, November 2020.

[AND 20c] ANDRA, Inventaire national des matières et déchets radioactifs. Les essentiels, Report, ANDRA, 2020.

[ANG 76] ANGELO J.A., Nuclear waste management by in situ melting, PhD, Department of Nuclear Engineering, University of Arizona, Tucson, AZ; University Microfilms, Ann Arbor, MI, USA, 1976.

[ANO 88] ANONYMOUS, "Mini-dossier. Le rapport Goguel. Stockage des déchets radioactifs en formations géologiques. Critères techniques du choix de sit", *CEN de Cadarache*, Supplément au Bulletin Mensuel no. 64, September 1988.

[ANO 20a] ANONYMOUS, Transmutation, *La radioactivité.com*, available at: https://www.laradioactivite.com/site/pages/TransmutationNucleaire.htm, 2020.

[ANO 20b] ANONYMOUS, Suède et Finlande : une gestion considérée comme modèle, available at: https://www.laradioactivite.com/site/pages/GestionSuede.htm#:~:text=On%20y%20stockera%209%20000,et%20des%20réacteurs%20en%20projet, 2020.

[APT 17] APTED M.J., AHN J., "Repository 101: Multiple-barrier geological repository design and isolation strategies for safe disposal of radioactive materials", in APTED M.J., AHN J. (eds), *Geological Repository Systems for Safe Disposal of Spent Nuclear Fuels and Radioactive Waste*, 2nd ed., Elsevier, Sawston 2017.

[ARE 15] ARENTSEN M.J., "Nuclear waste governance in the Netherlands", in BRUNNENGRÄBER A., DI NUCCI M.R., LOSADA A.M.I. *et al.* (eds), *Nuclear Waste Governance: An International Comparison*, Springer VS, Wiesbaden, 2015.

[ASN 06] ASN, Avis de l'Autorité de sûreté nucléaire sur les recherches relatives à la gestion des déchets à haute activité et à vie longue (HAVL) menées dans le cadre de la loi du 30 décembre 1991, et liens avec le PNGDR-MV, Report, ASN, 2006.

[ASN 08] ASN, Décision no. 2008-DC-0111 de l'Autorité de sûreté nucléaire du 2 septembre 2008 relative à la reprise et au conditionnement des boues actuellement entreposées dans l'atelier STE 2 (IBN 38), available at: https://www.asn.fr/content/download/52699/364306/version/2/file/Decision-2008-DC-0111.pdf, 2008.

[ASN 13] ASN, Avis no. 2013-AV-0187 de l'Autorité de sûreté nucléaire du 4 juillet 2013 sur la transmutation des éléments radioactifs à vie longue, Report, ASN, 2013.

[ASN 14] ASN, Rapport de l'ASN sur l'état de la sûreté nucléaire et de la radioprotection en France en 2013, Report, ASN, 2014.

[ASN 18a] ASN, Avis no. 2018-AV-0300 de l'Autorité de sûreté nucléaire du 11 janvier 2018 relatif au dossier d'options de sûreté présenté par l'Andra pour le projet Cigéo de stockage de déchets radioactifs en couche géologique profonde, Report, ASN, 2018.

[ASN 18b] ASN, Avis no. 2018-AV-0316 de l'Autorité de sûreté nucléaire du 18 octobre 2018 relatif à la cohérence du cycle du combustible nucléaire en France, Report, ASN, CODEP-DRC-2018-041575, 2018.

[ASN 18c] ASN, Avis relatif au dossier d'options de sûreté du projet de piscine d'entreposage centralisé d'assemblages combustibles usés d'EDF, Report, Groupe permanent d'experts pour les laboratoires et les usines (GPU), December 20, 2018.

[ASN 19a] ASN, Avis sur le dossier d'options de sûreté du projet de piscine d'entreposage centralisé de combustibles usés, Report, ASN, Avis no. 2019-AV-0331, Paris, July 23, 2019.

[ASN 19b] ASN, Étude PNGMDR 2016-2018 : comportement physico-chimique et thermique des colis de déchets bitumés en stockage, Report, ASN, CODEP-DRC-2019-005828, 2019.

[ASN 19c] ASN, Études PNGMDR – gestion à long terme des déchets radioactifs déjà produits du site de Malvési, Report, ASN, 2019.

[ASN 19d] ASN, Rapport de l'ASN sur l'état de la sûreté nucléaire et de la radioprotection en France en 2018, Report, ASN, 2019.

[ASN 19e] ASN, Revue externe sur la gestion des déchets bitumés. Rapport final, Report, ASN, June 28, 2019.

[ASN 20a] ASN, Avis no. 2020-AV-0357 de l'Autorité de sûreté nucléaire du 6 août 2020 sur les études relatives à la gestion des déchets de faible activité à vie longue (FA-VL) remises en application du plan national de gestion des matières et des déchets radioactifs 2016-2018, en vue de l'élaboration du cinquième plan national de gestion des matières et des déchets radioactifs, Report, ASN, 2020.

[ASN 20b] ASN, Avis sur les études concernant la gestion des matières radioactives et l'évaluation de leur caractère valorisable remises en application du plan national de gestion des matières et des déchets radioactifs 2016-2018, en vue de l'élaboration du cinquième plan national de gestion des matières et des déchets radioactifs, Report, ASN, Notice no. 2020-AV-0363 of the Autorité de sûreté nucléaire, October 8, 2020.

[AUF 15] AUFFERMANN B., SUOMELA P., KAIVO-OJA J. *et al.*, "A final solution for a big challenge, The governance of nuclear waste disposal in Finland", in MEZ L., DI NUCCI M.R., BRUNNENGRABER A. *et al.* (eds), *Governance of Nuclear Waste Management: An International Comparison*, Springer VS, Wiesbaden, 2015.

[BAI 91] BAIRD J.R., Subductive waste disposal method, US Patent number 5,022,788, June 11, 1991.

[BAI 05] BAIRD J., Disposal of high-level material in subduction zones – A rebuttal, available at: www.nwmo.ca/~/media/Site/Reports/2015/11/11/06/32/1287_baird-submission onthetopic_cho.ashx, 2005.

[BAL 14] BALA A., "Sub-seabed burial of nuclear waste: If the disposal method could succeed technically. Could it also succeed legally?", *Boston College Environmental Affairs Law Review*, vol. 41, no. 2, Article 6, 2014.

[BAN 19] BANERJEE P., HAZRA A., GHOSH P. *et al.*, "Solid waste management in India: A brief review", in GHOSH S.K. (ed.), *Waste Management and Resource Efficiency*, Springer Nature Singapore Pte Ltd., Singapore, 2019.

[BAR 09] BARTHE Y., "Les qualités politiques des technologies. Irréversibilité et réversibilité dans la gestion des déchets nucléaires", *Tracés*, vol. 1, no. 16, pp. 119–137, 2009.

[BAR 10] BARTHE Y., CALLON M., LASCOUMES P., "De la décision politique réversible : histoire d'une contribution inattendue de l'industrie nucléaire (française) a l'instauration de la démocratie dialogique", *Brazilian Journal of Urban Management*, vol. 1, no. 2, pp. 57–70, 2010.

[BAT 90] BATAILLE C., Rapport sut la gestion des déchets nucléaires à haute activité, Report no. 1839, Assemblée nationale, December 14, 1990.

[BAT 97] BATAILLE C., Rapport sur l'évolution de la recherche sur la gestion des déchets nucléaires haute activité – Les déchets militaires, Report to the Assemblée nationale no. 541, Office parlementaire d'évaluation des choix scientifiques et technologiques, December 15, 1997.

[BEL 17] BELMOKHTAR M., Contributions à l'étude du comportement thermo-hydro-mécanique de l'argilite du Callovo-Oxfordien (France) et de l'argile à Opalinus (Suisse), Thesis, Université Paris-Est, 2017.

[BEN 10] BENSENOUCI F., Apport des traceurs naturels à la compréhension des transferts au sein des formations argileuses compactées, Thesis, Université de Paris-Sud XI, 2010.

[BEN 13] BENSENOUCI F., MICHELOT J.L., MATRAY J.M. *et al.*, "Profiles of chloride and stable isotopes in pore-water obtained from a 2000 m-deep borehole through the Mesozoic sedimentary series in the eastern Paris Basin", *Physics and Chemistry of the Earth, Parts A/B/C*, vol. 65, p. 1610, 2013.

[BER 15] BERGMANS A., SUNDQVIST G., KOS D. *et al.*, "The participatory turn in radioactive waste management: Deliberation and the social-technical divide", *Journal of Risk Research*, vol. 18, no 3, pp. 347–363, 2015.

[BER 18] BERNARD E., LOTHENBACH B., CAN-DIT-COUMES C. *et al.*, "Magnesium and calcium silicate hydrates, Part I: Investigation of the possible magnesium incorporation in calcium silicate hydrate (C-S-H) and of the calcium in magnesium silicate hydrate (M-S-H)", *Applied Geochemistry*, vol. 89, pp. 229–242, 2018.

[BIS 74] BISHOP W.P., HOLLISTER C.D., "Seabed disposal: Where to look", *Nuclear Technology*, vol. 24, no. 3, pp. 425–443, 1974.

[BLA 01] BLACKMAN Jr. W.C., *Basic Hazardous Waste Management*, 3rd ed., Lewis Publishers, Boca Raton, 2001.

[BLA 16] BLANCK J., "Gouverner par le temps. Cadrages temporels du problème des déchets radioactifs et construction d'une irréversibilité technique", *Gouvernement et action publique*, vol. 1, no. 1, pp. 91–116, 2016.

[BLA 17] BLANCK J., Gouverner par le temps. La gestion des déchets radioactifs en France, entre changements organisationnels et construction de solutions techniques irréversibles (1950–2014), Thesis, Sciences Po-Institut d'études politiques de Paris, 2017.

[BLE 17] BLECHSCHMIDT I., VOMVORIS S., "Relevance of underground rock laboratories for deep geological repository programs", in APTED M.J., AHN J. (eds), *Geological Repository Systems for Safe Disposal of Spent Nuclear Fuels and Radioactive Waste*, 2nd ed., Elsevier, Sawston, 2017.

[BOI 00] BOISSON P., HUET P., MINGASSON J., Mission collégiale de concertation Granite, Article 1 of decree no. 99–687 of August 3, 1999, Report to the French Minister of Territory and Environment, to the Minister of Research, to the Secretary of State attached to the Minister of the Economy, Finance and Industry, in charge of industry, June 2000.

[BOI 11] BOISSIER F., "L'optimisation des filières de gestion des déchets radioactifs : un enjeu pour l'avenir", *Contrôle*, vol. 190, pp. 95–97, 2011.

[BON 11] BONIN B., *Les déchets nucléaires : état des lieux et perspectives*, EDP Sciences, Les Ulis, 2011.

[BOS 70] BOSTROM R.C., SHERIFF M.A., "Disposal of waste material in tectonic sinks", *Nature*, vol. 228, pp. 154–156, October 10, 1970.

[BOU 86] BOUST D., Les terres rares au cours de la diagenèse des sédiments abyssaux : analogies avec un transuranien : l'américium, Thesis, Université de Caen, 1986.

[BOU 97] BOUST D., COLIN C., LECLERC G. *et al.*, "Distribution and transit times of plutonium-bearing particles throughout the Channel", *Radioprotection*, no. 32, Special Issue Radionuclides in the Oceans, 1997.

[BOU 07] BOUST D., CHARMASSON S., "Le comportement des radionucléides dans les cycles biogéochimiques : le rôle des sédiments", in PEDRO G. (ed.), *Cycles biogéochimiques et écosystèmes continentaux*, Académie des Sciences, EDP, Les Ulis, 2007.

[BOU 10] BOUISSOU J., En Inde, la filière du recyclage échappe encore à tout contrôle, *Le Monde*, available at: https://www.lemonde.fr/planete/article/2010/04/28/en-inde-la-filiere-du-recyclage-echappe-encore-a-tout-controle_1343928_3244.html, April 28, 2010.

[BRA 12] BRADY P., ARNOLD B., ALTMAN S. *et al.*, Deep borehole disposal of nuclear waste: Final report, Report SAND2012-7789, Sandia, September 2012.

[BRA 17] BRADY P.V., FREEZE G.A., KUHLMAN K.L. *et al.*, "Deep borehole disposal of nuclear waste: US perspective", in APTED M.J., AHN J. (eds), *Geological Repository Systems for Safe Disposal of Spent Nuclear Fuels and Radioactive Waste*, 2nd ed., Elsevier, Sawston, 2017.

[BRE 78] BREDEHOEFT J.D., ENGLAND A.W., STEWART D.B. *et al.*, Geologic disposal of high-level radioactive wastes- earth-science perspectives, Circular 779, U.S. Geological Survey, 1978.

[BUD 18] BUDZIER A., FLYVBJERG B., GARAVAGLIA A. *et al.*, Quantitative cost and schedule risk analysis of nuclear waste storage, *Oxford Global Projects*, available at: https://arxiv.org/ftp/arxiv/papers/1901/1901.11123.pdf, December 10, 2018.

[BUR 78] BURNS R.F., CAUSEY W.E., GALLOWAY W.E. *et al.*, Nuclear waste disposal in space, Technical Paper 1225, NASA, May 1978.

[BUR 15] BURSÍK M., "The trouble with democracy: The challenges of nuclear waste governance in the Czech Republic", in BRUNNENGRÄBER A., DI NUCCI M.R., LOSADA A.M.I. *et al.* (eds), *Nuclear Waste Governance: An International Comparison*, Springer VS, Wiesbaden, 2015.

[BUS 18] BUSER M., WILDI W., "Du stockage des déchets toxiques dans des dépôts géologiques profonds", *Science et pseudo-sciences*, vol. 324, pp. 33–41, 2018.

[CAI 08] CAILLETEAU C., ANGELI F., DEVREUX F. *et al.*, "Insight into silicate-glass corrosion mechanisms", *Nature Materials*, vol. 7, pp. 978–983, December 2008.

[CAL 89] CALMET D.P., "L'immersion des déchets radioactifs dans l'océan : le point sur la question", *AIEA Bulletin*, pp. 47–50, 1989.

[CAR 15] CARLINO V., "S'opposer à Cigéo (Bure) : informer ou contre-informer ?", *4èmes journées doctorales sur la participation et la démocratie participative*, GIS Démocratie et Participation, Lille, November 2015.

[CAR 17] CARLIOZ T., Nucléation et propagation de fissures en conditions anisotropes, Thesis, Université Paris-Est, 2017.

[CAS 82] CASTAING R., Rapport du groupe de travail sur la gestion des combustibles irradiés, Report, Conseil supérieur de la sûreté nucléaire, November 1982.

[CAS 02] CASSAGNOU Y., DUBOIS M.A., "Le stockage géologique des déchets nucléaires", in BONCHE P. (ed.), *Le nucléaire expliqué par des physiciens*. EDP Sciences, Les Ulis, 2002.

[CAU 09] CAURANT D., LOISEAU P., MAJERUS O. *et al.*, *Glasses, Glass-Ceramics and Ceramics for Immobilization of Highly Radioactive Nuclear Wastes*, Nova Science Publishers, Inc., New York, 2009.

[CAV 02] CAVEDON J.M., "L'entreposage de longue durée", in BONCHE P. (ed.), *Le nucléaire expliquée par des physiciens*. EDP Sciences, Les Ulis, 2002.

[CCS 17] CCSN, La science derrière le stockage sûr des déchets nucléaires : des décennies de recherche en réglementation, available at: http://nuclearsafety.gc.ca/fra/pdfs/DGR/DGR-chronology-full-graphic-fra.pdf, 2017.

[CCS 20] CCSN, Déchets des mines et usines de concentration d'uranium, available at: https://www.nuclearsafety.gc.ca/fra/waste/uranium-mines-and-millswaste/index.cfm, 2020.

[CEA 08] CEA, Le conditionnement des déchets nucléaires, Monograph, CEA, DEN, Editions Le Moniteur, 2008.

[CEA 12] CEA, Séparation-transmutation des éléments radioactifs à vie longue, Chapter 2, CEA, DEN, available at: https://www.cea.fr/multimedia/Documents/publications/rapports/rapport-gestion-durable-matieres-nucleaires/Tome%202.pdf, December 2012.

[CEA 15] CEA, Avancées des recherches sur la séparation-transmutation et le multi-recyclage du plutonium dans les réacteurs à flux de neutrons rapides, Report, CEA, June 2015.

[CEA 17] CEA, L'assainissement-démantèlement des installations nucléaires, Monograph, CEA, DEN, Editions Le Moniteur, 2017.

[CEA 18a] CEA, Etat d'avancement des travaux sur des scénarios prospectifs de traitement et conditionnement des déchets d'enrobés bitumineux, MOA/Cab.AG/18-119 report, CEA, June 29, 2018.

[CEA 18b] CEA, Le Mémento sur l'énergie, édition 2018, Report, CEA, 2018.

[CEA 19] CEA, Rapport environnemental 2018. Bilans relatifs aux rejets gazeux et liquides des installations du CEA INBS et ICPE sur le site de Cadarache, Report, CEA, 2019.

[CHE 11] CHEVET P.F., "La gouvernance en matière de gestion des déchets radioactifs – la loi du 28 juin 2006", *Contrôle*, vol. 190, pp. 20–22, 2011.

[CHE 13] CHEN W., HAO J., CHEN Z., "A study of self-burial of a radioactive waste container by deep rock melting", *Science and Technology of Nuclear Installations*, Article ID 184757, p. 6, 2013.

[CLA 17] CLARK K.J., BERRYMAN K.R., "Assessing long-term stability of the geological environment", in APTED M.J., AHN J. (eds), *Geological Repository Systems for Safe Disposal of Spent Nuclear Fuels and Radioactive Waste*, 2nd ed., Elsevier, Sawston, 2017.

[CND 19a] CNDP, "Débat public. Plan national de gestion des matières et des déchets radioactifs", 5th edition 2019–2021. Report of the public debate between 17 April and 25 September 2019 drawn up by the President of the special committee for public debate, November 25, 2019.

[CND 19b] CNDP, Q20, producteurs de déchets radioactifs pour le médical, Report, CNDP, cndp-QUES-2019-04-127-Q20, April 23, 2019.

[CNE 18] CNE2, Des recherches et études relatives à la gestion des matières et des déchets radioactifs, Evaluation report no. 12, CNE2, June 2018.

[CNE 19a] CNE2, Des recherches et études relatives à la gestion des matières et des déchets radioactifs, Evaluation report no. 13, CNE2, June 2019.

[CNE 19b] CNE2, Stockage des colis de déchets bitumes. Note d'information au parlement sur le rapport final de la revue externe sur la gestion des déchets bitumes, Note, CNE2, October 2019.

[CNE 20] CNE2, Des recherches et études relatives à la gestion des matières et des déchets radioactifs, Evaluation report no. 14, CNE2, June 2020.

[COL 93] COLE Jr. H.W., Collapsible waste disposal container and method of disposal of waste in subduction zone between tectonic plates, US Patent 5 245 118, September 14, 1993.

[COU 19a] COUR DES COMPTES, L'aval du cycle du combustible nucléaire. Les matières et les déchets radioactifs, de la sortie du réacteur au stockage, Press release, Cour des Comptes, July 4, 2019.

[COU 19b] COUR DES COMPTES, L'aval du cycle du combustible nucléaire. Les matières et les déchets radioactifs, de la sortie du réacteur au stockage, Public thematic report, Cour des Comptes, July 4, 2019.

[CRO 17] CROSSLAND I., "Surface, subsurface, intermediate depth, and borehole disposal", in APTED M.J., AHN J. (eds), *Geological Repository Systems for Safe Disposal of Spent Nuclear Fuels and Radioactive Waste*, 2nd ed., Elsevier, Sawston, 2017.

[DAH 09] DAHLE S., SAVINOV V., CARROLL J. *et al.*, "A return to the nuclear waste dumping sites in the bays of Novaya Zemlya", *Radioprotection*, vol. 44, no. 5, pp. 281–284, 2009.

[DAY 04] DAY D.E., RAY C.S., KIM C.W., Iron phosphate glasses: An alternative for vitrifying certain nuclear wastes, Final report for DE-FG07-96ER45618, Project Number: 73976, University of Missouri-Rolla, Rolla, 2004.

[DEF 08] DEFRA, Managing radioactive waste safely, A framework for implementing geological disposal, Report, Defra, Cm 7386, June 2008.

[DEG 12] DEGTEVA M.O., SHAGINA N.B., VOROBIOVA M.I. *et al.*, "Reevaluation of waterborne releases of radioactive materials from the Mayak Production Association into the Techa River in 1949–1951", *Health Physics*, vol. 102, no. 1, pp. 25–38, 2012.

[DEL 14] DELAY J., BOSSART P., LING L.X. *et al.*, "Three decades of underground research laboratories: What have we learned?" in NORRIS S., BRUNO J., CATHELINEAU M. *et al.* (eds), *Clays in Natural and Engineered Barriers for Radioactive Waste Confinement*, Geological Society, London, Special Publications, March 5, 2014.

[DEL 16] DELAGE P., MENACEUR H., BELMOKHTAR M. *et al.*, "Effet d'une saturation partielle sur la réponse hydromécanique d'une roche argileuse", *Journal of Applied Science, Engineering and Technology*, vol. 1, no. 2, pp. 9–13, 2016.

[DEN 16] DENOUN M., "La gestion des déchets nucléaires : un cas de disjonction entre science et progrès?", *Mouvements*, vol. 87, no. 3, pp. 13–24, 2016.

[DEP 02] DE PRETER P., LALIEUX P., "Fulfilment of the long-term safety functions by the different barriers during the main time frames after repository closure", in NEA (ed.), "The handling of timescales in assessing post-closure safety of deep geological repositories", *Workshop Proceedings*, Paris, France, 16–18 April 2002.

[DEP 03] DEPARTMENT OF EDUCATION, SCIENCE AND TRAINING, "Rehabilitation of former nuclear test sites at emu and Maralinga (Australia) 2003", *Maralinga Rehabilitation Technical Advisory Committee*, Commonwealth of Australia, 2003.

[DEP 15] DEPAUS C., SKAGIUS K., BUCKAU G. *et al.*, GEOSAF II: Task group on operational safety. Progress Report, IAEA, Vienna, available at: https://www.iaea.org/sites/default/files/19/02/geosaf-2-report_task_group.pdf, 2015.

[DES 18] DESSUS B., "Belleville sur Loire, la conséquence de la faillite de la doctrine du retraitement", *Le Club Mediapart*, Monday, February 19, 2018.

[DIN 15] DI NUCCI M.R., "Breaking the stalemate: The challenge of nuclear waste governance in Italy", in BRUNNENGRÄBER A., DI NUCCI M.R., LOSADA A.M.I. *et al.* (eds), *Nuclear Waste Governance: An International Comparison*, Springer VS, Wiesbaden, 2015.

[DOA 17] DOAN P.H.L., Prise en compte économique du long terme dans les choix énergétiques relatifs à la gestion des déchets radioactifs, Thesis, Economy and Finance, PSL Research University, Paris, 2017.

[DOE 17] DoE, Sixth National Report for the Joint Convention on the Safety of Spent Fuel Management and on the Safety of Radioactive Waste Management. Report, U.S. Department of Energy, USA, October 2017.

[DSN 10] DSND, Rapport du DSND sur le suivi radioécologique des INBS et gestion des anciens sites d'entreposage des déchets radioactifs, available at: https://inis.iaea.org/collection/NCLCollectionStore/_Public/42/024/42024447.pdf, 2010.

[EDF 04] EDF, Nucléaire et Environnement 2002–2003, Report, EDF, France, 2004.

[EDF 16] EDF, Rapport de l'Inspecteur Général pour la Sûreté Nucléaire et la Radioprotection, Report, EDF, France, 2016.

[EVR 11] EVRARD L., "La gestion des déchets radioactifs : avancées et perspectives", *Contrôle*, vol. 190, pp. 6–11, 2011.

[EWI 95] EWING R.C., WEBERT W.J., CLINARD Jr. F.W. "Radiation effects in nuclear waste forms for high-level radioactive waste", *Progress in Nuclear Energy*, vol. 29, no. 2, pp. 63–121, 1995.

[EWI 16] EWING R.C., WHITTLESTON R.A., YARDLEY B.W.D., "Geological disposal of nuclear waste: A primer", *Elements*, vol. 4, no. 12, pp. 233–237, 2016.

[FAU 19] FAUVEL B., MARX S., "La gestion et le stockage radioactifs", *Report and Note of the CESER Grand Est*, June 13, 2019.

[FIG 18] FIG D., "Disposal and contamination nuclear waste governance in South Africa", in BRUNNENGRÄBER A., DI NUCCI M.R., LOSADA A.M.I. *et al.* (eds), *Challenges of Nuclear Waste Governance: An International Comparison*, Volume II, Springer VS, Wiesbaden, 2018.

[FIS 16] FISHER A.J., CORKHILL C.L., HAND R.J. *et al.*, "Intrinsic dissolution rate determination of vitrified high level waste", in VARDON P.J., BYKOV D. (eds), *Proceedings of the 2nd Petrus-OPERA PhD and Early Stage Researcher Conference*, Delft University of Technology, June 27, 2016.

[FON 60] FONTAINE Y., La contamination radioactive des milieux et des organismes aquatiques, CEA-R-1589 report, 1960.

[FOR 15] FORREST R.A., "'Yucca mountain is dead': The challenge of nuclear waste governance in the United States", in BRUNNENGRÄBER A., DI NUCCI M.R., LOSADA A.M.I. *et al.* (eds), *Nuclear Waste Governance: An International Comparison*, Springer VS, Wiesbaden, 2015.

[FRE 19] FREEZE G.A., STEIN E., BRADY P.V. *et al.*, "Deep borehole disposal safety case", *Energies*, vol. 12, p. 2141, 2019.

[FRE 01] FREMERY L., RIGAUD S., *Que deviennent les effluents radioactifs hospitaliers ?*, ISIM, Sciences et technologie de l'eau, Montpellier, January 2001.

[FRO 11] FROMONOT C., RANCHER J., DOUCHE C. *et al.*, "L'entreposage des déchets solides tritiés sans filière", *Contrôle*, vol. 190, pp. 106–102, 2011.

[GAR 14] GARCIER R., "Disperser, confiner ou recycler ?", *L'Espace géographique*, vol. 43, no. 3, pp. 265–283, 2014.

[GAR 18] GARZIA E., "Lutter pour le territoire, vivre le territoire. Le cas du site d'enfouissement des déchets radioactifs à Bure. CIST2018-Représenter les territoires/Representing territories", *CIST*, Rouen, France, March 2018.

[GAS 18] GASPAR M., MAYHEW N., URAM 2018 : les hauts et les bas – les paramètres économiques de l'extraction d'uranium, *Bulletin de l'AIEA*, available at: https://www.iaea.org/fr/newscenter/news/uram-2018-les-hauts-et-les-bas-les-parametres-economiques-de-lextraction-duranium, June 2018.

[GIB 99] GIBB F.G.F., "High-temperature, very deep, geological disposal: A safer alternative for high-level radioactive waste?" *Waste Management*, vol. 19, pp. 207–211, 1999.

[GOG 87] GOGUEL J., Stockage des déchets radioactifs en formations géologiques. Critères techniques de choix de site, Report of the Working Group presided over by Professeur Goguel, Ministry of Industry, P.&T. and Tourism, June 1985–May 1987.

[GOL 18] GOLOVKINA A., OVSYANNIKOV D., OLARU S., "Performance optimization of radioactive waste transmutation in accelerator driven system", *Cybernetics and Physics*, vol. 4, no. 7, pp. 210–215, 2018.

[GOU 04] GOUZY A., Étude du comportement du plutonium au cours de la diagenèse précoce des sédiments marin : applications à deux environnements marins marqués par des rejets de radionucléides issus d'usines de retraitement, Thesis, Université de Caen Basse-Normandie, December 2004.

[GOU 05] GOUZY A., BOUST D., CONNAN O. *et al.*, "Diagenetic reactivity of the plutonium in marine anoxic sediments (Cumbrian mud patch – eastern Irish Sea)", *Radioprotection*, vol. 40, no. 1, pp. S541–S547, 2005.

[GRI 02] GRIFFAULT L., FILLION E., "Handling of timescales and related safety indicators", in NEA (ed.), "The handling of timescales in assessing post-closure safety of deep geological repositories", *Workshop Proceedings Paris*, France, April 16–18, 2002.

[GUE 12] GUENELEY P., MATHIEU J.P., STENGER M. *et al.*, "Liminaire", in BERTAUX F., BERTHET, J. *et al.* (eds), *Gestion des déchets nucléaires, réflexion et questions sur les enjeux éthiques*, Meuse Nature Environnement, Bar-le-Duc, November 2012.

[GUE 15] GUEROULT R., HOBBS D.T., FISCH N.J., "Plasma filtering techniques for nuclear waste remediation", *Journal of Hazardous Materials*, arXiv:1504.07180v1 [physics.plasm-ph], April 27, 2015.

[GUI 16a] GUILLEMETTE A., ZERBIB J.C., AUTRET J.C., "Quid du retour vers leur pays d'origine des déchets issus du retraitement des combustibles de réacteurs à eau légère usés 'importés' ?", *Les cahiers de Global Chance*, no. 38, pp. 56–70, January 2016.

[GUI 16b] GUILLEMETTE A., ZERBIB J.C., AUTRET J.C., "La réalité du recyclage des combustibles usés des REP d'EDF de 1976 à 2012", *Les cahiers de Global Chance*, no. 38, pp. 71–78, January 2016.

[GUO 20] GUO X., GIN S., LEI P. *et al.*, "Self-accelerated corrosion of nuclear waste forms at material interfaces", *Nature Materials*, vol. 19, pp. 310–316, 2020.

[HCT 18] HCTISN, "Présentation du 'Cycle du combustible' français en 2018", *HCTISN*, September 21, 2018.

[HCT 20a] HCTSIN, "Avis no. 13 et recommandations du Haut comité sur les perspectives d'évolution de la filière de gestion des déchets très faiblement radioactifs (TFA) en France", *HCTSIN*, April 7, 2020.

[HCT 20b] HCTSIN, "Perspectives d'évolutions de la filière de gestion des déchets très faiblement radioactifs (TFA). Rapport final", *HCTSIN*, April 7, 2020.

[HEU 81] HEUZE F.E., "On the geotechnical modeling of high-level nuclear waste disposal by rock melting", *Lawrence Livermore Laboratory*, Livermore, CA, UCRL-S3183, 1981.

[HIC 80] HICKOX C.E., GARTLING D.K., MCVEY D.F. *et al.*, "Analysis of heat and mass transfer in sub-seabed disposal of nuclear waste", *Marine Geotechnology*, vol. 5, pp. 3–4, 1980.

[HOC 15a] HOCKE P., KALLENBACH-HERBERT B., "Always the same old story? Nuclear waste governance in Germany", in BRUNNENGRÄBER A., DI NUCCI M.R., LOSADA A.M.I. *et al.* (eds), *Nuclear Waste Governance: An International Comparison*, Springer VS, Wiesbaden, 2015.

[HOC 15b] HOCKE P., KUPPLER S., "Participation under tricky conditions: The Swiss nuclear waste strategy based on the sectoral plan", in BRUNNENGRÄBER A., DI NUCCI M.R., LOSADA A.M.I. *et al.* (eds), *Nuclear Waste Governance: An International Comparison*, Springer VS, Wiesbaden, 2015.

[HOL 81] HOLLISTER C.D., ANDERSON D.R., HEATH G.R., "Subseabed disposal of nuclear wastes", *Science*, vol. 213, no. 1321, pp. 1321–1326, 1981.

[IAE 99] IAEA, Inventory of radioactive waste disposals at sea, Report, TECDOC 1105, IAEA, Vienna, 1999.

[IAE 00] IAEA, Management of radioactive waste from the use of radionuclides in medicine, Report, IAEA-TECDOC-1183, IAEA, Vienna, 2000.

[IAE 03] IAEA, "The long term storage of radioactive waste: Safety and sustainability. A position paper of international experts", *Technical Committee Meetings*, Vienna, Austria 2–5 April 2002; 20–22 November 2002, IAEA-LTS/RW, IAEA, Vienna, 2003.

[IAE 04] IAEA, Application of the concepts of exclusion. Exemption and clearance, Safety guide, IAEA Safety Standards Series No. RS-G-1.7, 2004.

[IAE 09a] IAEA, Classification of radioactive waste: General safety guide, Guide, IAEA Safety Standards Series No GSG-1, IAEA, Vienna, 2009.

[IAE 09b] IAEA, Predisposal management of radioactive waste, Guide, IAEA Safety Standards Series No. GSR Part 5, IAEA, Vienna, 2009.

[IAE 11a] IAEA, Disposal of radioactive waste: Specific safety requirements, Report, IAEA Safety Standards Series No. SSR-5, Publication 1449, IAEA, Vienna, 2011.

[IAE 11b] IAEA, GEOSAF I the international Intercomparison and harmonisation project on demonstrating the safety of geological disposal, Final Report, IAEA, Vienna, available at: https://www.iaea.org/sites/default/files/19/02/geosaf-i-final-report.pdf, 2011.

[IAE 12a] IAEA, Storage of spent nuclear fuel. Specific safety guide, Guide, IAEA Safety Standards Series no. SSG-15, Publication 1503, IAEA, Vienna, 2012.

[IAE 12b] IAEA, The safety case and safety assessment for the disposal of radioactive waste. Specific safety guide, Guide, IAEA Safety Standards Series No. SSG-23, Publication 1553, IAEA, Vienna, 2012.

[IAE 15] IAEA, Managing integration of pre-closure activities and post-closure safety in the safety case for geological disposal, IAEA, Vienna, available at: https://www.iaea.org/sites/default/files/19/02/geosaf-2-tecdoc-draft.pdf, 2015.

[IAE 16a] IAEA, Processing of irradiated graphite to meet acceptance criteria for waste disposal. Results of a coordinated research project, Report, IAEA-TECDOC-1790, IAEA, Vienna, 2016.

[IAE 16b] IAEA, International conference on the safety of radioactive waste management, Report, IAEA-CN-242, Vienna, November 21–25, 2016.

[IAE 16c] IAEA, International peer review on the "Safety options dossier" of the project of disposal of radioactive waste in deep geological formations: Cigéo, Final report, IAEA, Vienna, 2016.

[IAE 17a] IAEA, Selection of technical solutions for the management of radioactive waste, Report, IAEA-TECDOC-1817, Publication TE-1817, IAEA, Vienna, 2017.

[IAE 17b] IAEA, Management of large volumes of waste arising in a nuclear or radiological emergency, Report, IAEA-TECDOC-1826, Publication TE-1826, IAEA, Vienna, 2017.

[IAE 17c] IAEA, GEOSAF Part III; The international intercomparison and harmonisation project on demonstration of the operational and long-term safety of geological disposal facilities for radioactive waste. Terms of reference, Report, IAEA, Vienna, available at: https://www.iaea.org/sites/default/files/19/02/geosaf-3-draft-tor.pdf, 2017.

[IAE 18a] IAEA, Status and trends in spent fuel and radioactive waste management, Report, IAEA Nuclear Energy Series, no. NW-T-1.14, Publication 1799, IAEA, Vienna, 2018.

[IAE 18b] IAEA, A methodology for establishing a national strategy for education and training in radiation. Transport and waste safety, Report, Safety Reports Series, no. 93, Publication 1778, IAEA, Vienna, 2018.

[IAE 19a] IAEA, Predisposal management of radioactive waste from the use of radioactive material in medicine: Industry, agriculture, research and education, IAEA safety standards for protecting people and the environment, Specific Safety Guide, no. SSG-45, Publication 1758, Vienna, 2019.

[IAE 19b] IAEA, Demonstrating performance of spent fuel and related storage system components during very long term storage. Final report of a coordinated research project, Report, IAEA-TECDOC-1878, Publication TE-1878, Vienna, 2019.

[IAE 19c] IAEA, "Management of spent fuel from nuclear power reactors. An integrated approach to the back end of the fuel cycle", *Proceedings of an International Conference*, IAEA, Vienna, Austria, 15–19 June 2015, Publication STIPUB 1850, 2019.

[IAE 19d] IAEA, Methodology for the systematic assessment of the regulatory competence needs (SARCoN) for regulatory bodies of radiation facilities and activities, Report, IAEA TECDOC-1860, Publication TE-1860, Vienna, 2019.

[IAE 19e] IAEA, Nuclear fuel cycle simulation system: Improvements and applications, Reports, IAEA-TECDOC-1864, Publication TE-1864, Vienna, 2019.

[IAE 20] IAEA, "Management of spent fuel from nuclear power reactors. Learning from the past, enabling the future", *Proceedings of an international conference*, IAEA, Vienna, 24–28 June 2019, IAEA Publication P1905, Vienna, 2020.

[IEE 11] IEER, Examen critique du programme de l'Andra sur les recherches effectuées dans le laboratoire souterrain de bure et sur la zone de transposition pour définir une ZIRA, Final report, IEER, April 20, 2011.

[IRS 13a] IRSN, Les déchets radioactifs. Définitions, classement et modes de gestion, available at: https://www.irsn.fr/dechets/dechets-radioactifs/Pages/Definitions-classement-gestion.aspx, 2013.

[IRS 13b] IRSN, La gestion des déchets radioactifs. Collecter, trier, traiter, conditionner, entreposer et finalement stocker pour se protéger, IRSN, available at: https://www.irsn.fr/dechets/dechets-radioactifs/Documents/irsn_livret_dechets_radioactifs.pdf, 2013.

[IRS 13c] IRSN, La gestion pour les déchets à vie longue, available at: https://www.irsn.fr/dechets/dechets-radioactifs/Pages/gestion-dechets-vie-longue.aspx, 2013.

[IRS 13d] IRSN, Le stockage en couche géologique profonde à l'international, available at: https://www.irsn.fr/dechets/dechets-radioactifs/Pages/stockage-international.aspx, 2013.

[IRS 13e] IRSN, La séparation/transmutation des déchets à vie longue. Thème 1 – La place de Cigéo dans le dispositif de gestion des déchets, Public debate on radioactive waste, IRSN, 2013.

[IRS 13f] IRSN, "Le stockage géologique : option incontournable du dispositif de gestion durable des déchets radioactifs ?", *Cahier d'acteurs*, no. 1, May 2013.

[IRS 16] IRSN, Accidents survenus au WIPP en février 2014 : synthèse des actions effectuées, available at: https://www.irsn.fr/dechets/actualite/Pages/20160727_Accidents-survenus-au-WIPP-en-fevrier-2014.aspx, July 27, 2016.

[IRS 17] IRSN, Projet de stockage cigéo. Examen du dossier d'options de sûreté, Report, IRSN no. 2017-00013, 2017.

[IRS 18a] IRSN, Entreposage du combustible nucléaire usé : concepts et enjeux de sûreté, IRSN report no. 2018-00003, 2018.

[IRS 18b] IRSN, Méthodologie et critères envisageables pour apprécier la nocivité des matières et déchets radioactifs, Report in response to Article 1 of the Order of February 23, 2017 establishing the requirements of the national radioactive materials and waste management plan, Report no. PSE-ENV/2018-00048, IRSN, Pôle santé et environnement, 2018.

[IRS 18c] IRSN, Le transport de matières radioactives, available at: https://www.irsn.fr/FR/connaissances/Installations_nucleaires/transport/transports-matieres-radioactives/Pages/0-transports-matieres-radioactives-sommaire.aspx#.X3M8bu3grQw, March 2018.

[IRS 18d] IRSN, Piscine d'entreposage centralisé d'assemblages combustibles usés d'EDF. Examen du dossier d'options de sûreté, Report, IRSN, December 6, 2018.

[IRS 19a] IRSN, Analyse des possibilités d'entreposage à sec de combustibles nucléaires usés de type MOX ou URE, IRSN report no. 2019-00265, IRSN, 2019.

[IRS 19b] IRSN, Fiches pédagogiques de synthèse sur les recherches de l'IRSN no. 1 à 10, Report, IRSN, 2019.

[IRS 19c] IRSN, Panorama international des recherches sur les alternatives au stockage géologique des déchets de haute et moyenne activité à vie longue, Report, IRSN/2019-00318 report, May 2019.

[IVE 12] IVERSON N., PERSON M., "Glacier-bed geomorphic processes and hydrologic conditions relevant to nuclear waste disposal", *Geofluids*, no. 12, pp. 38–57, 2012.

[JAI 18] JAITNER F., "A profitable business strategy? Spent nuclear fuel and radioactive waste management in Russia", in BRUNNENGRÄBER A., DI NUCCI M.R., LOSADA A.M.I. *et al.* (eds), *Challenges of Nuclear Waste Governance: An International Comparison*, Springer VS, vol. 2, Wiesbaden, 2018.

[JIM 18] JIMENO M., "A democratic deficit: The challenges of nuclear waste governance in Argentina and Brazil", in BRUNNENGRÄBER A., DI NUCCI M.R., LOSADA A.M.I. *et al.* (eds), *Challenges of Nuclear Waste Governance: An International Comparison*, Springer VS, vol. 2, Wiesbaden, 2018.

[JOR 69] JORF, "Décret du 19 juin 1969 autorisant la création d'une installation pour le stockage de déchets radioactifs solides", *Journal officiel de la République française*, June 22, 1969.

[JOR 92a] JORF, "Loi no. 91-1381 du 30 décembre 1991 relative aux recherches sur la gestion des déchets radioactifs", *Journal officiel de la République française*, p. 10, January 1, 1992.

[JOR 92b] JORF, "Arrêté du 17 décembre 1992 portant nomination d'un médiateur", *Journal officiel de la République française*, p. 17347, December 17, 1992.

[JOR 99] JORF, "Arrêté du 19 novembre 1999 portant nomination à une mission de concertation", *Journal officiel de la République française*, p. 17674, November 27, 1999.

[JOR 06a] JORF, "LOI no 2006-686 du 13 juin 2006 relative à la transparence et à la sécurité en matière nucléaire", *Journal officiel de la République française*, June 14, 2006.

[JOR 06b] JORF, "LOI no 2006-739 du 28 juin 2006 de programme relative à la gestion durable des matières et déchets radioactifs", *Journal officiel de la République française*, June 29, 2006.

[JOR 11] JORF, "Décret no. 2011-1910 du 20 décembre 2011 autorisant l'Agence nationale pour la gestion des déchets radioactifs à exploiter sur le territoire de la commune de Bure (Meuse) un laboratoire souterrain destiné à étudier les formations géologiques profondes où pourraient être stockés des déchets radioactifs", *Journal officiel de la République française*, December 22, 2011.

[JOR 14] JORF, "Arrêté du 4 avril 2014 modifiant l'arrêté du 9 octobre 2008 relatif à la nature des informations que les responsables d'activités nucléaires et les entreprises mentionnées à l'article L. 1333-10 du code de la santé publique ont obligation d'établir, de tenir à jour et de transmettre périodiquement à l'Agence nationale pour la gestion des déchets radioactifs", *Journal officiel de la République française*, April 27, 2014.

[JOR 15a] JORF, "Décret no. 2015-231 du 27 février 2015 relatif à la gestion des sources radioactives scellées usagées", *Journal officiel de la République française*, March 1, 2015.

[JOR 15b] JORF, "Arrêté du 4 mai 2015 créant une commission d'information auprès des anciens sites d'expérimentations nucléaires du Pacifique", *Journal officiel de la République française*, May 20, 2015.

[JOR 16] JORF, "Arrêté du 11 janvier 2016 portant homologation de la décision no 2015-DC-0536 de l'Autorité de sûreté nucléaire du 22 décembre 2015 fixant les valeurs limites de rejet dans l'environnement des effluents liquides et gazeux des installations nucléaires de base nos 33 (UP2-400), 38 (STE2 et AT1), 47 (ELAN II B), 80 (HAO), 116 (UP3-A), 117 (UP2-800) et 118 (station de traitement des effluents STE3) exploitées par AREVA NC sur le site de La Hague (département de la Manche)", *Journal officiel de la République française*, January 14, 2016.

[JOR 17] JORF, "Décret no. 2017-231 du 23 février 2017 pris pour application de l'article L. 542-1-2 du code de l'environnement et établissant les prescriptions du plan national de gestion des matières et des déchets radioactifs", *Journal officiel de la République française*, February 23, 2017.

[JOU 11] JOUE, "Directive 2011/70/EURATOM du conseil du 19 juillet 2011 établissant un cadre communautaire pour la gestion responsable et sûre du combustible usé et des déchets radioactifs", *Journal officiel de l'Union européenne* of August 2, 2011, L199/48 to L199/56, 2011.

[KÅB 15] KÅBERGER T., SWAHN J., "Model or muddle? Governance and management of radioactive waste in Sweden", in BRUNNENGRÄBER A., DI NUCCI M.R., LOSADA A.M.I. *et al.* (eds), *Nuclear Waste Governance: An International Comparison*, Springer VS, Wiesbaden, 2015.

[KES 17] KESSLER J.H., "Effects of very long-term interim storage of spent nuclear fuel and HLW on subsequent geological disposal", in APTED M.J., AHN J. (eds), *Geological Repository Systems for Safe Disposal of Spent Nuclear Fuels and Radioactive Waste*, 2nd ed., Elsevier, Sawston, 2017.

[KIC 97] KICKMAIER W., MCKINLEY I., "A review of research carried out in European rock laboratories", *Nuclear Engineering and Design*, no. 176, pp 75–81, 1997.

[KLE 74] KLETT R.D., Deep rock nuclear waste disposal test: Design and operation, Report, SAN93-2723, UC-721, Sandia National Laboratories, Albuquerque, NM, 1974.

[KLE 97] KLETT R.D., Performance assessment overview subseabed disposal of high-level radioactive waste, Sandia Report, SAN93-2723, UC-721, Albuquerque, NM, 1997.

[KOR 18] KORITÁR Z., "Postponed policy: Nuclear waste governance in Hungary", in BRUNNENGRÄBER A., DI NUCCI M.R., LOSADA A.M.I. *et al.* (eds), *Challenges of Nuclear Waste Governance: An International Comparison*, Springer VS, vol. 2, Wiesbaden, 2018.

[KRU 79] KRUTENAT R.A., Method for the disposal of nuclear or toxic waste materials, US Patent 4,178,109, December 11, 1979.

[LAB 11] LABALETTE T., "Le projet HA-MAVL : vers la réalisation du centre industriel de stockage géologique Cigéo", *Contrôle*, vol. 190, pp. 98–102, 2011.

[LAL 94] LALLEMAND S.E., SCHNURLE P., MALAVIEILLE J., "Coulomb theory applied to accretionary and non-accretionary wedges: Possible causes for tectonic erosion and/or frontal accretion", *Journal of Geophysical Research*, vol. 99, no. B6, pp. 12033–12055, 1994.

[LAL 16] LALAN P., DAUZÈRES A., DE WINDT L. *et al.*, "Impact of a 70 C temperature on an ordinary Portland cement paste/claystone interface: An *in situ* experiment", *Cement and Concrete Research*, vol. 83, pp. 164–178, 2016.

[LAP 17] LAPONCHE B., ZERBIB J.C., La filière des surgénérateurs au Japon – Monju, Tokaï et Rokkashô, Global Chance, available at: http://www.global-chance.org/IMG/pdf/bljcz3_japon_-_monju_et_rokkasho.pdf, October 16, 2017.

[LAP 18] LAPONCHE B., "L'entreposage à sec des combustibles nucléaires irradiés : pas si bête...", Global Chance, May 5, 2018.

[LAR 00] LARSON R.W., "Development of the conceptual models for chemical conditions and hydrology used in the 1996 performance assessment for the waste isolation pilot plant", *Reliabil Eng System Saf*, vol. 69, pp. 59–86, 2000.

[LAV 16] LAVEROV N.P., YUDINTSEV S.V., KOCHKIN B.T. *et al.*, "The Russian strategy of using crystalline rock as a repository for nuclear waste", *Elements*, vol. 12, pp. 253–256, 2016.

[LEE 18] LEEM S.J., SCHREURS M., "Transitioning away from opaque governance nuclear waste policy in South Korea", in BRUNNENGRÄBER A., DI NUCCI M.R., LOSADA A.M.I. *et al.* (eds), *Challenges of Nuclear Waste Governance: An International Comparison*, Springer VS, vol. 2, Wiesbaden, 2018.

[LOC 00] LOCHARD J., SCHIEBER C., SCHNEIDER T. *et al.*, La gestion du risque associé aux déchets radioactifs à haute activité et à vie longue, Report no. 268, CEPN, 2000.

[LOS 15] LOSADA A.M.I., "Subject to political capture? Nuclear waste governance in Spain", in BRUNNENGRÄBER A., DI NUCCI M.R., LOSADA A.M.I. *et al.* (eds), *Nuclear Waste Governance: An International Comparison*, Springer VS, Wiesbaden, 2015.

[LUC 04] LUCEY JA., GOUZY A., BOUST D. *et al.*, "Geochemical fractionation of plutonium in anoxic Irish sea sediments using an optimised sequential extraction protocol", *Applied Radiation and Isotopes*, no. 60, pp. 379–385, 2004.

[MAC 96] MACILWAIN C., "Russian waste goes to ground", *Nature*, vol. 383, p. 376, 1996.

[MAC 15] MACKERRON G., "Multiple challenges nuclear waste governance in the United Kingdom", in BRUNNENGRÄBER A., DI NUCCI M.R., LOSADA A.M.I. *et al.* (eds), *Nuclear Waste Governance: An International Comparison*, Springer VS, Wiesbaden, 2015.

[MAC 19] MACHADO G.G., GUILHEN S.N., KRUPSKAYA V.V. *et al.*, "Brazilian clays as potential buffer materials for radioactive waste final storage", *Associação Brasileira de Energia Nuclear (ABEN), 2019 International Nuclear Atlantic Conference – INAC 2019*, Santos, SP, Brazil, October 21–25, 2019.

[MAH 17] MAHJOUB M., Modélisation du couplage endommagement-perméabilité dans les géomatériaux anisotropes : application aux ouvrages souterrains du site de Bure, Thesis, Université de recherche Paris Sciences et Lettres, 2017.

[MAK 00] MAKHIJANI A., ZERRIFFI H., The nuclear alchemy gamble: An assessment of transmutation as a nuclear waste management strategy, Technical report, IEER, 2000.

[MAR 97] MARSILLY G.D., "Enfouissement des déchets nucléaires en formation géologique", in TURLAY R. (ed.), *Les déchets nucléaires, un dossier scientifique*, EDP Sciences, Les Ulis, 1997.

[MAR 07] MARTIN G., Les atolls de Mururoa et de Fangataufa (Polynésie française), Les expérimentations nucléaires, Aspects radiologiques, CEA-R-6136 report, 2007.

[MCD 01] MCDONALD P., VIVES I., BATTLE J. *et al.*, "The availability of plutonium and americium in Irish Sea sediments for redissolution", *The Science of the Total Environment*, no. 267, pp. 109–123, 2001.

[MEA 07] MEALE A., Les déchets radioactifs et la protection de l'environnement, Doc. 11377, Report by the Commission de l'environnement, de l'agriculture et des questions territoriales, European Union, available at: http://assembly.coe.int/nw/xml/XRef/X2H-Xref-ViewHTML.asp?FileID=11772&lang=fr, September 17, 2007.

[MER 16] MERCIER-BION F., LEON Y., NEFF D. *et al.*, "Effet des bactéries sur la corrosion d'un acier en contexte nucléaire : caractérisation par une approche multi-techniques (microscopie optique, µRaman et MEB-EDS)", *Matériaux & Techniques*, vol. 104, no. 511, 2016.

[MER 18] MERCAT C., LAMOUROUX C., "Un changement drastique dans le stockage des déchets nucléaires de faibles activités ?", *Radioprotection*, vol. 53, no. 4, pp. 249–254, 2018.

[MIH 18] MIHÓK P., "Delays in finding a solution: The governance of nuclear waste disposal in Slovakia", in BRUNNENGRÄBER A., DI NUCCI M.R., LOSADA A.M.I. *et al.* (eds), *Challenges of Nuclear Waste Governance: An International Comparison*, Springer VS, vol. 2, Wiesbaden, 2018.

[MIL 14] MILES E.L., "Sub-seabed disposal of high level radioactive waste: The policy context then and now", *The Oceans in the Nuclear Age: Legacies and Risks: Expanded Edition*, pp. 125–143, 2014.

[MIN 15] MINISTÈRE DE LA DÉFENSE, "Surveillance des atolls de Mururoa et de Fangataufa, Tome I, Surveillance radiologique", Year 2013, no. 09, DEF/DGA/DO/UM NBC/SCEN of January 30, 2015.

[MTE 18] MTES (Ministère de la transition écologique et solidaire), Rapport national de la France de mise en œuvre de la Directive du 19 juillet 2011, Report, MTES, 2018 edition.

[MUR 06] MURITH C., ZELLER W., HAMMANS M. *et al.*, "Réexamen de la justification du tritium dans l'industrie horlogère", *Radioprotection*, vol. 41, no. 2, pp. 189–200, 2006.

[MUR 16] MURITH C., BAECHLER S., ESTIER S. *et al.*, "Remediation of radium legacies from de Swiss watch industry", *Radiation Protection Dosimetry*, vol. 173, nos 1–3, pp. 245–251, 2016.

[NAD 96] NADIS S., "The sub-seabed solution.", *The Atlantic Monthly*, vol. 278, no. 4, pp. 28–39, available at: www.stoppingclimatechange.com/Barge%20-%20Sub-Seabed%20Waste%20Solution%20-%2096.10.pdf, October 1996.

[NAK 15] NAKAJIMA K., *Nuclear Back-End and Transmutation Technology for Waste Disposal: Beyond the Fukushima Accident*, Springer Open, Tokyo, 2015.

[NEA 96] NEA, Co-ordinated research and environmental surveillance programme related to sea disposal of radioactive waste, CRESP Final report 1981–1995. NEA, OECD, 1996.

[NEA 01] NEA, The role of underground laboratories in nuclear waste disposal programmes, Report Radioactive waste management, NEA, OECD, Paris, 2001.

[NEA 02] NEA, "The handling of time-scales in assessing post-closure safety of deep geological repositories", *Workshop Proceedings*, OECD, NEA, Paris, April 16–18, 2002.

[NEA 10] NEA, "Towards transparent, proportionate and deliverable regulation for geological disposal", *Workshop Proceedings*, Tokyo, Japan, 20–22 January 2009, NEA no. 6825, OECD, Paris, 2010.

[NEA 13] NEA, "Underground research laboratories (URL)", *Radioactive Waste Management*, NEA/RWM/R(2013)2, NEA no. 78122, OECD, Paris, February 2013.

[NEA 15a] NEA, Stakeholder involvement in decision-making: A short guide to issues, Approaches and resources, Report, NEA, 2015.

[NEA 15b] NEA, "Radioactive waste management and constructing memory for future generations", *Proceedings of the International, Conference and Debate*, 15–17 September 2014, NEA, Verdun, France, 2015.

[NEA 15c] NEA, "Fostering a durable relationship between a waste management facility and its host community", *Adding Value through Design and Process*, NEA, 2015.

[NEA 16a] NEA, "Management of radioactive waste after a nuclear power plant accident", *NEA*, no. 7305, OECD, Paris, 2016.

[NEA 16b] NEA, "National inventories and management strategies for spent nuclear fuel and radioactive waste. Methodology for common presentation of data", *NEA*, no. 7323, OECD, Paris, 2016.

[NEA 16c] NEA, Japan's siting process for the geological disposal of high-level radioactive waste, an international peer review, Report, NEA, 2016.

[NEA 17a] NEA, Communication on the safety case for a deep geological repository, Report, NEA, 2017.

[NEA 17b] NEA, Sourcebook of international activities related to the development of safety cases for deep geological repositories, Report, NEA, 2017.

[NEA 17c] NEA, "International conference on geological repositories 2016", *Conference Synthesis*, 7–9 December 2016, Paris, France, NEA, 2017.

[NEA 17d] NEA, National inventories and management strategies for spent nuclear fuel and radioactive waste. Extended methodology for the common presentation of data, Report, NEA, 2017.

[NEA 18a] NEA, Metadata for radioactive waste management, Report, NEA, 2018.

[NEA 18b] NEA, Microbial influence on the performance of subsurface, salt-based radioactive waste repositories. An evaluation based on microbial ecology, bioenergetics and projected repository conditions, Report, NEA, 2018.

[NEA 19] NEA, "Preservation of records, knowledge and memory (RK&M) across generations. developing a key information file for a radioactive waste repository", *NEA*, no. 7421, Paris, 2019.

[NII 15] NIIBORI Y., "Radioactive waste disposal", in NAGASAKI S., NAKAYAMA S., (eds), *Radioactive Waste Engineering and Management*, Springer, Tokyo, 2015.

[NIR 04] NIREX, A review of the deep borehole disposal concept for radioactive waste, Nirex Report no. N/108, United Kingdom Nirex Limited, June 2004.

[NRC 15] NRC, NRC regulations title 10: Code of federal regulations, Part 61.55 Waste classification, available at: www.nrc.gov/reading-rm/doccollections/cfr/part061/part061-0055.html, December 2, 2015.

[NRC 16] NRC (Nuclear Regulatory Commission), Dry cask storage of spent nuclear fuel, Backgrounder, Office of Public Affairs, available at: http://www.nrc.gov/reading-rm/doc-collections/fact-sheets/dry-cask-storage.html, 2016.

[NYF 03] NYFFELER N., CIGNA A.A., DAHLGAARD H. *et al.*, "Radionuclides in the Atlantic Ocean: A survey", in GUÉGUÉNIAT P., GERMAIN P.P., MÉTIVIER H. (eds), *Radionuclides in the Oceans, Inputs and Inventories*, EDP Sciences, Les Ulis, 2003.

[OEC 88] OECD, *Feasibility of Disposal of High-Level Radioactive Waste into the Seabed. Overview of Research and Conclusions*, vol. 1, OECD, Paris, 1988.

[OEC 18] OECD, "State-of-the-art report on the progress of nuclear fuel cycle chemistry", *Nuclear Science, NEA*, no. 7267, OECD, Paris, 2018.

[OJO 05] OJOVAN M.I., LEE W.E., *An Introduction to Nuclear Waste Immobilisation*, Elsevier, Amsterdam, 2005.

[OJO 12] OJOVAN M., POLUEKTOV P., KASCHEEV V., "Self-disposal option for highly-radioactive waste reconsidered", *Nuclear Engineering International*, pp. 28–29, July 2012.

[OJO 14] OJOVAN M.I., LEE W.E., *An Introduction to Nuclear Waste Immobilisation*, 2nd ed., Elsevier, Amsterdam, 2014.

[OKA 18] OKAMURA L., "False premise, false promise governance and management of nuclear waste in Japan", in BRUNNENGRÄBER A., DI NUCCI M.R., LOSADA A.M.I. *et al.* (eds), *Challenges of Nuclear Waste Governance: An International Comparison*, Springer VS, vol. 2, Wiesbaden, 2018.

[OKO 15] OKOSHI M., MOMMA T., "Radioactive waste treatment technologies", in NAGASAKI S., NAKAYAMA S. (eds), *Radioactive Waste Engineering and Management*, Springer, Tokyo, 2015.

[OND 01] ONDRAF/NIRAS, SAFIR 2, Safety assessment and feasibility interim report 2, Report, NIROND 2001-06E, 2001.

[ONE 02] O'NEILL K., "Radioactive 'trade': Globalizing the nuclear fuel cycle", *SAIS Review*, vol. 22, no. 1, pp. 157–168, Winter–Spring 2002.

[ONW 83] ONWI, Very deep hole systems engineering studies, Report ONWI-226, US Office of Nuclear Waste Isolation, Battelle Laboratories, Columbus, OH, 1983.

[ORA 18a] ORANO, Extended optimized storage: Dry shielded canister, available at: http://us.areva.com/EN/home-3138/areva-nuclear-materials-tn-americas--nuhoms-used-fuel-storage-system.html, 2018.

[ORA 18b] ORANO, Traitement des combustibles usés provenant de l'étranger dans les installations d'Orano la Hague. Rapport 2017, Report, Orano, June 29, 2018.

[ORA 19] ORANO, Les résidus du traitement de conversion de l'uranium (RTCU) : Orano Cycle – Malvési, PNGMDR public debate, Atelier thématique de Narbonne, June 6, 2019.

[ORE 18] ORELLANA L.F., SCUDERI M.M., COLLETTINI C. *et al.*, "Frictional properties of opalinus clay: Implications for nuclear waste storage", *Journal of Geophysical Research: Solid Earth*, vol. 123, pp. 157–175, 2018.

[OSB 18] OSBORN D.M., Discussion of the Yucca Mountain Project and doe spent fuel disposition, Sandia national laboratories, SAND2018-3922C, available at: https://www.osti.gov/servlets/purl/1507829, 2018.

[PAP 01] PAPP T., "Topical session on the policy basis for successful stepwise decision making in repository development", *14th Meeting of the OECD NEA, Radioactive Waste Management Committee*, Paris, March 2001.

[PAR 18] PAROTTE C., L'art de gouverner les déchets hautement radioactifs. Analyse comparée de la Belgique, la France et le Canada, Thesis, Université de Liège, Presses Universitaires de Liège, 2018.

[PAS 18] PASYUK O., "Too much to handle. Radioactive waste management in a post nuclear accident state: Ukraine", in BRUNNENGRÄBER A., DI NUCCI M.R., LOSADA A.M.I. *et al.* (eds), *Challenges of Nuclear Waste Governance: An International Comparison*, Springer VS, vol. 2, Wiesbaden, 2018.

[PAT 02] PATARIN L., *Le cycle du combustible nucléaire*, EDP Sciences, Les Ulis, 2002.

[PAT 18] PATINAUX L., "Enfouir des déchets nucléaires dans un monde conflictuel. Une histoire de la démonstration de sûreté de projets de stockage géologique, en France (1982–2013)", *EHESS*, 2018.

[PAT 19] PATINAUX L., "Enjeux épistémiques et politiques des recherches sur l'évacuation géologique des déchets nucléaires. Étude d'une controverse sur l'implantation d'un laboratoire souterrain dans la Vienne (1994–1998)", *Cahiers François Viète*, Centre François Viète, Université de Nantes, pp. 133–157, 2019.

[PAU 13] PAUL M., "Experience from the WISMUT project with the management of large volumes of radioactive residues", *International Experts Meeting on Decommissioning and Remediation after a Nuclear Accident*, IAEA, Vienna, January 28–February 1, 2013.

[PEY 92] PEYRIN J.O., "La gestion des déchets radioactifs hospitaliers", *Radioprotection*, vol. 27, no. 1, pp. 47–53, 1992.

[PHI 77] PHILBERTH K., "The disposal of radioactive waste in ice sheets", *Journal of Glaciology*, vol. 81, no. 19, pp. 607–617, 1977.

[PNG 10] PNGMDR, Plan national de gestion des matières et des déchets radioactifs 2010–2012, Report, Ministère de l'écologie, de l'Energie, du développement durable et de la mer, Direction générale de l'énergie et du climat, 2010.

[PNG 17] PNGMDR, Sixième rapport national sur la mise en œuvre des obligations de la Convention commune, Report, Convention commune sur la sûreté de la gestion du combustible usé et sur la sûreté de la gestion des déchets radioactifs, PNGMDR, October 2017.

[PNG 20] PNGMDR, Avis de la commission 'Orientations' sur la note de la maîtrise d'ouvrage relative à la gestion des déchets FA-VL, Report, PNGMDR, December 21, 2020.

[POI 13] POISSON R., "France: Experience of radioactive waste (RAW) management and contaminated site clean-up", in LEE W.E., OJOVAN M.I., JANTZEN C.M. (eds), "Radioactive waste management and contaminated site clean-up: Processes, technologies and international experience", *Woodhead Publishing Series in Energy*, no. 48, Woodhead Publishing, Oxford, 2013.

[POS 18] POSKAS P., "Progress on nuclear waste management nuclear waste governance in Lithuania", in BRUNNENGRÄBER A., DI NUCCI M.R., LOSADA A.M.I. *et al.* (eds), *Challenges of Nuclear Waste Governance: An International Comparison*, Springer VS, vol. 2, Wiesbaden, 2018.

[QUI 83] QUINN R.E., ROHRBAUGH J.W., "A spatial model of effectiveness criteria towards a competing values approach to organizational effectiveness", *Management Science*, no. 29, pp. 363–377, 1983.

[RAH 15] RAHMAN R.O.A., RAKHIMOV R.Z., RAKHIMOVA N.R. *et al.*, *Cementitious Materials for Nuclear Waste Immobilization*, John Wiley & Sons, Chichester, 2015.

[RAN 02] RANFT M., WOLLRATH J., "Timescales in the long-term safety assessment of the Morsleben repository, Germany", in NEA (ed.) "The handling of timescales in assessing post-closure safety of deep geological repositories", *Workshop Proceedings*, Paris, France, April 16–18, 2002.

[REA 10] REAUD C., SCHIEBER C., SCHNEIDER T. *et al.*, Stockage géologique de déchets radioactifs : mise en œuvre pratique du concept de réversibilité et gouvernance, Report no. 308, CEPN (Centre d'étude sur l'évaluation de la protection dans le domaine nucléaire), 2010.

[REV 06] REVOL H., Projet de loi de programme relatif à la gestion durable des matières et des déchets radioactifs, Report no. 358 (2005–2006), Made on behalf of the committee on economic affairs, May 17, 2006.

[RIC 78] RICE A.L., "Radio-active waste disposal and deep-sea biology", *Oceanologica Acta*, vol. 1, no. 4, pp. 483–491, 1978.

[RIC 81] RICE E.E., PRIEST C.C., "An overview of nuclear waste disposal in space" in HOFMANN P.L., BRESLIN J.J. (eds), *The Technology of High-Level Nuclear Waste Disposal Advances in the Science and Engineering of the Management of High-Level Nuclear Wastes*, vol. 1, DOE/TIC-4621, 1981.

[ROC 18] ROCHEFORT D., "Définir une norme pour l'éternité ? L'exemple des messages d'avertissement autour des sites de stockage de déchets nucléaires", in LIQUETE V., MALLOWAN M., MARCON C. (eds), *Processus de normalisation et durabilité de l'information*, *Revue COSSI*, no. 5, pp. 116–128, 2018.

[ROD 17] RODRÍGUEZ-PENALONGA L., MORATILLA SORIA B.Y., "A review of the nuclear fuel cycle strategies and the spent nuclear fuel management technologies", *Energies*, no. 10, p. 1235, 2017.

[RUI 04] RUI S., CHATAIGNIER S., PIERLOT S., "Le débat impensable ? L'enfouissement des déchets nucléaires au risque de la démocratie", in CASTAGNA B., GALLAIS S., RICAUD P. *et al.* (eds), *La situation délibérative dans le débat public*, Presses Universitaires François-Rabelais, 2004.

[SAF 18] SAFONOV A.V., ZAKHAROVA E.V., NAZINA T.N. *et al.*, "Russian experience on microbiological investigation of groundwater from the zone of liquid radioactive waste deep burial", *Radioactive Waste*, vol. 3, no. 4, pp. 32–40, 2018.

[SAM 16] SAMPIETRO D., ABARCA E., BAYER M. *et al.*, "Long-term evaluation on the groundwater chemistry due to cement materials with numerical simulation", in VARDON P.J., BYKOV D. (eds), *Proceedings of the 2nd Petrus-OPERA PhD and Early Stage Researcher Conference*, Delft University of Technology, June 27, 2016.

[SAN 16] SANDERS M.C., SANDERS C.E., "A world's dilemma 'upon which the sun never sets' – The nuclear waste management strategy: Western European nation states and the United States of America, Part I", *Atw, International Journal of Nuclear Power*, vol. 11, no. 61, pp. 667–673, November, 2016.

[SAN 17] SANDERS M.C., SANDERS C.E., "A world's dilemma 'upon which the sun never sets' – The nuclear waste management strategy: Western European nation states and the United States of America, Part III", *Atw, International Journal of Nuclear Power*, vol. 1, no. 62, pp. 35–45, January 2017.

[SAW 00] SAWHILL S.G., "Cleaning-up the arctic's cold war legacy: Nuclear waste and Arctic military environmental cooperation", *Cooperation and Conflict*, vol. 35, no. 1, pp. 5–36, 2000.

[SAW 01] SAWHILL S.G., JØRGENSEN A.K., Military nuclear waste and international cooperation in northwest Russia, Report 12, FNI, 2001.

[SCH 74a] SCHNEIDER K.J., PLATT A.M., High-level radioactive waste management alternatives, Report, Volume 1, Section 1: Summary, Section 2: Background and Data Base, Section 3: Evaluation Methodology, Battelle Pacific Northwest Laboratories, BNWL-1900, May 1974.

[SCH 74b] SCHNEIDER K.J., PLATT A.M., High-level radioactive waste management alternatives, Report, Volume 2, Section 4: Geological Disposal, Battelle Pacific Northwest Laboratories, BNWL-1900, May 1974.

[SCH 74c] SCHNEIDER K.J., PLATT A.M., High-level radioactive waste management alternatives, Report, Volume 3, Section 5: Ice Sheet Disposal, Section 6: Seabed Disposal, Battelle Pacific Northwest Laboratories, BNWL-1900, May 1974.

[SCH 74d] SCHNEIDER K.J., PLATT A.M., High-level radioactive waste management alternatives, Report, Volume 4, Section 7: Waste partitioning, Section 8: Extraterrestrial disposal, Section 9: Transmutation processing, Battelle Pacific Northwest Laboratories, BNWL-1900, May 1974.

[SCH 74e] SCHNEIDER, K.J., PLATT, A.M., High-level radioactive waste management alternatives, Report, Battelle Mem. Inst., Pacific Northwest Labs. [Rept.] BNWL-1900, 4 v., 1974.

[SCH 78] SCHWARTZ L.L., COHEN J.J., LEWIS A.E. *et al.*, High-level radioactive waste isolation by incorporation in silicate rocks: Peaceful Nuclear Explosions V, Report, International Atomic Energy Agency, Vienna, IAEA-TC-81-5/4, 1978.

[SCH 15] SCHRÖDER J., BERGMANS A., LAES E., "Advanced research: Lagging policy nuclear waste governance in Belgium", in BRUNNENGRÄBER A., DI NUCCI M.R., LOSADA A.M.I. *et al.* (eds), *Nuclear Waste Governance: An International Comparison*, Springer VS, Wiesbaden, 2015.

[SCH 17] SCHWARTZ F.W., KIM Y., CHAE B.G., "Deep borehole disposal of nuclear wastes: Opportunities and challenges", *Journal of Nuclear Fuel Cycle and Waste Technology*, vol. 15, no. 4, pp. 301–312, December 2017.

[SEE 17] SEETHAPATHY S., HENDERSON J.H., "Management of construction waste in nuclear and thermal power plant projects in India", *Journal of Construction in Developing Countries*, vol. 22, supp. 1, pp. 19–46, 2017.

[SEN 17] SENGUPTA A., GUPTA N.K., "MWCNTs based sorbents for nuclear waste management: A review", *Journal of Environmental Chemical Engineering*, vol. 5, pp. 5099–5114, 2017.

[ŠEŠ 18] ŠEŠERKO L., "An arranged marriage nuclear waste governance and nuclear energy in Slovenia and Croatia", in BRUNNENGRÄBER A., DI NUCCI M.R., LOSADA A.M.I. *et al.* (eds), *Challenges of Nuclear Waste Governance: An International Comparison*, Springer VS, vol. 2, Wiesbaden, 2018.

[SEY 11] SEYVE C., Les rejets radioactifs des installations nucléaires (GR21 de la SFEN), available at: http://www.sfen.org/Les -rejets-radioactifs-des, 2011.

[SHR 16] SHRESTHA R., STEINOVÁ J., FALTEISEK L. *et al.*, "Characterization of microbial communities in raw and homogenized bentonite samples", in VARDON P.J., BYKOV D. (eds), *Proceedings of the 2nd Petrus-OPERA PhD and Early Stage Researcher Conference*, Delft University of Technology, June 27, 2016.

[SIL 72] SILVER E.A., "Subduction zones: Not relevant to present-day problems of waste disposal", *Nature*, vol. 239, pp. 330–331, October 6, 1972.

[SIL 06] SILVY J.P., "Comment rendre compatibles entreposage et longue durée ?", Nouveaux concepts : le colis au cœur des études, *Clefs CEA*, no. 53, pp. 64–81, Winter 2005–2006.

[SKB 18] SKB, Extending the SFR, available at: www.skb.se/publikation/2570733/Factsheet_Extending_the_SFR.pdf, January 2018.

[SLO 91] SLOVIC P., FLYNN J.H., LAYMAN M., "Perceived risk: Trust, and the politics of nuclear waste", *Science*, vol. 254, no. 5038, pp. 1603–1607, 1991.

[SMI 09] SMITH P., CORNÉLIS B., CAPOUET M. *et al.*, The long-term safety strategy for the geological disposal of radioactive waste, SFC1 level 4 report: Second full draft, ONDRAF/NIRAS, NIROND-TR 2009-12 E, June 2009.

[SOR 15] SORENSON K.B., "Safe and secure transport and storage of radioactive materials", *Woodhead Publishing Series in Energy*, Elsevier, Cambridge, no. 78, 2015.

[SOU 17] SOULEY M., KAZMIERCZAK J.B., Modélisation hydro-élasto-visco-plastique d'une galerie du laboratoire souterrain de Meuse/Haute-Marne de l'Andra, INERIS scientific report, 2017.

[STE 18a] STERNFELD E., "A long way off nuclear waste governance in China", in BRUNNENGRÄBER A., DI NUCCI M.R., LOSADA A.M.I. *et al.* (eds), *Challenges of Nuclear Waste Governance: An International Comparison*, Springer VS, vol. 2, Wiesbaden, 2018.

[STE 18b] STETTEN L., Spéciation et mobilité de l'uranium dans des sols et des sédiments lacustres en aval d'anciens sites miniers, Thesis defended in Paris, October 12, 2018.

[STE 18c] STETTEN L., BLANCHART P., MANGERET A. *et al.*, "Redox fluctuations and organic complexation govern uranium redistribution from U(IV)-phosphate minerals in a mining-polluted wetland soil, Brittany, France", *Environmental Science & Technology*, vol. 52, no. 22, pp. 13099–13109, 2018.

[STE 18d] STETTEN L., MANGERET A., BREST J. *et al.*, "Geochemical control on the reduction of U(VI) to mononuclear U(IV) species in lacustrine sediments", *Geochimica Cosmochimica Acta*, vol. 222, pp. 171–186, 2018.

[STO 02] STORCK R., BECKER D.A., "The spin project: Safety and performance indicators in different time frames", in NEA (ed.), "The handling of timescales in assessing post-closure safety of deep geological repositories", *Workshop Proceedings*, Paris, France, April 16–18, 2002.

[STU 17] STUK, Joint convention on the safety of spent fuel management and on the safety of radioactive waste management, 6th Finnish National Report, STUK-B 218, October 2017.

[TAC 15] TACHIBANA M., "Clearance", in NAGASAKI S., NAKAYAMA S. (eds), *Radioactive Waste Engineering and Management*, Springer, Tokyo, 2015.

[TAJ 17] TAJIMA T., MOUROU G., NAKAJIMA K., Free-electron laser driven by fiber laser-based laser plasma accelerator, US Patent US 9,768,580 B2, September 19, 2017.

[TAV 19] TAVARES B.L., TELLO C.C.O., "Concrete containers in radioactive waste management: A review", *Brazilian Journal of Radiation Sciences*, vol. 7, no. 2A, pp. 1–17, 2019.

[THA 15] THAKUR A., JADHAV M., CHAVAN N. *et al.*, "Radioactive waste management", *International Journal of Innovations in Engineering Research and Technology*, vol. 2, no. 5, pp. 1–8, May 2015.

[TIW 13] TIWARI Å.V., KADU P.A., "Biomedical waste management practices in India – A review", *International Journal of Current Engineering and Technology*, vol. 3, no. 5, pp. 2030–2033, December 2013.

[TON 06] TONHAUSER W., JANKOWITSCH-PREVOR O., "La convention commune sur la sûreté de la gestion du combustible usé et sur la sûreté de la gestion des déchets radioactifs", in AEN (ed.), *Le droit nucléaire international après Tchernobyl*, NEA no. 6147, OECD, Paris, 2006.

[TOS 17] TOSONI E., SALO A., ZIO E., "Scenario analysis for the safety assessment of nuclear waste repositories: A critical review", *Risk Analysis*, pp. 1–22, 2017.

[TOS 18] TOSHIKI T., MOUROU G., GALES S., Arrangement for generating a proton beam and an installation for transmutation of nuclear wastes, US Patent US 10,049,788 B2, August 14, 2018.

[TRA 93] TRAPEZNIKOV A.V., POZOTOTINA V.N., CHEBOTINA M. *et al.*, "Radioactive contamination of the Techa river: The Urals", *Health Physics*, vol. 65, no. 5, pp. 481–488, November 1993.

[TSA 05] TSANG C.F., BERNIER F., DAVIES C., "Geohydromechanical processes in the excavation damaged zone in crystalline rock, rock salt, and indurated and plastic clays – In the context of radioactive waste disposal", *International Journal of Rock Mechanics and Mining Sciences*, vol. 42, no. 1, pp. 109–125, 2005.

[UE 14] UE, "Directive 2013/59/Euratom du Conseil du 5 décembre 2013 fixant les normes de base relatives à la protection sanitaire contre les dangers résultant de l'exposition aux rayonnements ionisants et abrogeant les directives 89/618/Euratom, 90/641/Euratom, 96/29/Euratom, 97/43/Euratom et 2003/122/Euratom", *Journal Officiel de l'Union Européenne (JOUE)*, L 13, January 17, 2014.

[UME 02] UMEKI H., SMITH P.A., "Handling of timescales: Application of safety indicators", in OECD (ed.), "The handling of time-scales in assessing post-closure safety of deep geological repositories", *Workshop Proceedings*, Paris, April 16–18, 2002.

[UNI 10] UNITED NATIONS, "Sources and effects of ionizing radiation", *UNSCEAR 2008*, United Nations, New York, 2010.

[USA 74] USAEC, High-level radioactive waste management alternatives, Report, WASH-1297, May 1974.

[USD 16] USDoE, A historical review of the safe transport of spent nuclear fuel: Fuel cycle research and development, Report, FCRD-NFST-2016-000474, Rev. 1, August 31, 2016.

[USD 17] USDoE, Sixth national report for the joint convention on the safety of spent fuel management and on the safety of radioactive waste management, Report, U.S. Department of Energy, USA, October 2017.

[VES 18] VESTERGAARD C., "A twinned approach: The challenges of nuclear waste governance in Canada", in BRUNNENGRÄBER A., DI NUCCI M.R., LOSADA A.M.I. *et al.* (eds), *Challenges of Nuclear Waste Governance: An International Comparison*, Springer VS, vol. 2, Wiesbaden, 2018.

[VIR 17] VIRA J., "Geological repository for high-level nuclear waste becoming reality in Finland", in APTED M.J., AHN J. (eds), *Geological Repository Systems for Safe Disposal of Spent Nuclear Fuels and Radioactive Waste*, 2nd ed., Elsevier, Sawston, 2017.

[VOE 17] VOEGELE M.D., "The Yucca Mountain license application", in APTED M.J., AHN J. (eds), *Geological Repository Systems for Safe Disposal of Spent Nuclear Fuels and Radioactive Waste*, 2nd ed., Elsevier, Sawston, 2017.

[VON 17] VON BERLEPSCH T., "Salt repository systems: Design development approach at the example of the Gorleben salt dome", in APTED M.J., AHN J. (eds), *Geological Repository Systems for Safe Disposal of Spent Nuclear Fuels and Radioactive Waste*, 2nd ed., Elsevier, Sawston, 2017.

[WAT 13] WATTAL P.K., "Indian programme on radioactive waste management", *Sädhanä*, vol. 38, no. 5, pp. 849–857, October 2013.

[WNA 17a] WNA, Transport of radioactive materials, available at: https://www.world-nuclear.org/information-library/nuclear-fuel-cycle/transport-of-nuclear-materials/transport-of-radioactive-materials.aspx, 2017.

[WNA 17b] WNA, Japanese waste and MOX shipments from Europe, available at: https://www.world-nuclear.org/information-library/nuclear-fuel-cycle/transport-of-nuclear-materials/japanese-waste-and-mox-shipments-from-europe.aspx, 2017.

[WNN 17] WNN, "Urenco tails plant to start up in 2018", *World Nuclear News*, September 1, 2017.

[YAN 17] YAN X., YANG L., ZHANG X. *et al.*, "Concept of an accelerator-driven advanced nuclear energy system", *Energies*, vol. 10, no. 944, doi:10.3390/en10070944, 2017.

[YAN 19] YANG P., WANG Y., RODRIGUEZ M.A. *et al.*, "Rock-welding materials development for deep borehole nuclear waste disposal", *Materials Chemistry and Physics*, vol. 221, pp. 178–187, 2019.

[YOS 15] YOSHIDA A., "Considering the geological disposal program of high-level radioactive waste through classroom debate", in NAKAJIMA K. (eds), *Nuclear Back-end and Transmutation Technology for Waste Disposal. Beyond the Fukushima Accident*, Springer Open, Tokyo, 2015.

[YUA 17] YUAN H., Caractérisation expérimentale des propriétés de poro-mécaniques et de transfert de l'argilite du Cox, Thesis, Université de Lille, 2017.

[ZER 83] ZERBIB J.C., "Les éléments non techniques qui sous-tendent les choix", *Que faire des déchets radioactifs ?*, Journées scientifiques à l'Ecole des Mines de Paris, June 10–11, 1983.

[ZER 17] ZERBIB J.C., Retraitement des combustibles belges à La Hague et gestion des combustibles et des déchets radioactifs en Belgique, Working document, Global Chance, available at: http://www.global-chance.org/Retraitement-des-combustibles-belges-a-La-Hague-et-gestion-des-combustibles-et-des-dechets-radioactifs-en-Belgique, March 2017.

[ZER 18a] ZERBIB J.C., Usines de retraitement des combustibles irradiés de La Hague et leurs problèmes, Report to the commission d'enquête de l'assemblée nationale sur la sûreté et la sécurité nucléaires, Global Chance, May 31, 2018.

[ZER 18b] ZERBIB J.C., GUILLEMETTE A., "Plutonium entreposé à La Hague 2010–2016", Working document, Global Chance, February 13, 2018.

[ZER 20a] ZERBIB J.C., Les atteintes dues aux rayonnements X et au radium Personal communication, 2020.

[ZER 20b] ZERBIB J.C., "Questions sur les déchets de graphite FA-VL d'après le Catalogue descriptif des familles de l'ANDRA, 2018", Global Chance, December 29, 2020.

[ZHU 89] ZHU J.L., CHAN C.Y., "La gestion des déchets radioactifs dans le monde. Réalisations et perspectives actuelles", *AIEA Bulletin*, no. 4, pp. 5–13, 1989.

Index

A, C, D

F, G, H, I

L, M, N

P, R

Other titles from

ISTE

in

Ecological Science

2021

AMIARD Jean-Claude
Disarmament and Decommissioning in the Nuclear Domain
(Radioactive Risk SET – Volume 4)

JOLY Fernand, BOURRIÉ Guilhem
Mankind and Deserts 1: Deserts, Aridity, Exploration and Conquests
Mankind and Deserts 2: Water and Salts
Mankind and Deserts 3: Wind in Deserts and Civilizations

MUTTIN Frédéric, THOMAS Hélène
Marine Environmental Quality

2020

BRUSLÉ Jacques, QUIGNARD Jean-Pierre
Fish Behavior 1: Eco-ethology
Fish Behavior 2: Ethophysiology

LE FLOCH Stéphane
Remote Detection and Maritime Pollution: Chemical Spill Studies

MIGON Christophe, NIVAL Paul, SCIANDRA Antoine
The Mediterranean Sea in the Era of Global Change 1: 30 Years of Multidisciplinary Study of the Ligurian Sea
The Mediterranean Sea in the Era of Global Change 2: 30 Years of Multidisciplinary Study of the Ligurian Sea

ROSSIGNOL Jean-Yves
Climatic Impact of Activities: Methodological Guide for Analysis and Action

2019

AMIARD Jean-Claude
Industrial and Medical Nuclear Accidents: Environmental, Ecological, Health and Socio-economic Consequences
(Radioactive Risk SET – Volume 2)
Nuclear Accidents: Prevention and Management of an Accidental Crisis
(Radioactive Risk SET – Volume 3)

BOULEAU Gabrielle
Politicization of Ecological Issues: From Environmental Forms to Environmental Motives

DAVID Valérie
Statistics in Environmental Sciences

GIRAULT Yves
UNESCO Global Geoparks: Tension Between Territorial Development and Heritage Enhancement

KARA Mohamed Hichem, QUIGNARD Jean-Pierre
Fishes in Lagoons and Estuaries in the Mediterranean 2: Sedentary Fish
Fishes in Lagoons and Estuaries in the Mediterranean 3A: Migratory Fish
Fishes in Lagoons and Estuaries in the Mediterranean 3B: Migratory Fish

OUVRARD Benjamin, STENGER Anne
Incentives and Environmental Policies: From Theory to Empirical Novelties

2018

AMIARD Jean-Claude
Military Nuclear Accidents: Environmental, Ecological, Health and Socio-economic Consequences
(Radioactive Risk SET – Volume 1)

FLIPO Fabrice
The Coming Authoritarian Ecology

GUILLOUX Bleuenn
Marine Genetic Resources, R&D and the Law 1: Complex Objects of Use

KARA Mohamed Hichem, QUIGNARD Jean-Pierre
Fishes in Lagoons and Estuaries in the Mediterranean 1: Diversity, Bioecology and Exploitation

2016

BAGNÈRES Anne-Geneviève, HOSSAERT-MCKEY Martine
Chemical Ecology

2014

DE LARMINAT Philippe
Climate Change: Identification and Projections

Printed and bound by CPI Group (UK) Ltd, Croydon, CR0 4YY

15/09/2026

14962363-0001